Human Resource Implications of ROBOTICS

by

H. Allan Hunt

and

Timothy L. Hunt

The W. E. Upjohn Institute for Employment Research

Library of Congress Cataloging in Publication Data

Hunt, H. Allan.
Human resource implications of robotics.

Bibliography: p.
1. Unemployment, Technological—United States.
2. Automation—Social aspects—United States.
3. Automation—Economic aspects—United States.
I. Hunt, Timothy L. II. Title.
HD6331.2U5H86 1983 331.12'929892'0973 83-6 201
ISBN 0-88099-009-0
ISBN 0-88099-008-2 (pbk.)

300 South Westnedge Ave.
Kalamazoo, Michigan 49007

THE INSTITUTE, a nonprofit research organization, was established on July 1, 1945. It is an activity of the W. E. Upjohn Unemployment Trustee Corporation, which was formed in 1932 to administer a fund set aside by the late Dr. W. E. Upjohn for the purpose of carrying on "research into the causes and effects of unemployment and measures for the alleviation of unemployment."

Foreword

While there is considerable interest and concern about the use of robots in the workplace, most public awareness has been shaped by the popular press in the last year or two. In a context of serious concern about high levels of unemployment, there has been a growing need for thorough investigation and sound estimates and projections of the labor market effects of robotics. Nowhere is that need greater than in Michigan, where the auto industry is one of the nation's heaviest users of industrial robots.

This study was initiated at the request of the Michigan Occupational Information Coordinating Committee as an examination of the human resource implications of robotics for the State of Michigan. It was later expanded to focus on the impact of robots on the entire U.S. In the course of the study, many fears have appeared to be unfounded. There are also many areas of legitimate concern to human resource planners and policymakers who need to understand the implications of robotics for economic development, job creation, job displacement, training and retraining.

Facts and observations presented in this study are the sole responsibility of the authors. Their viewpoints do not necessarily represent the position of the W. E. Upjohn Institute for Employment Research.

Jack R. Woods
Acting Director

March 1983

Acknowledgments

We are deeply indebted to two individuals for the genesis of this study: John Koval of the Michigan Department of Labor and Robert Sherer of the Michigan Occupational Information Coordinating Committee. John first perceived the need for a Michigan-specific study of the human resource implications of robotics. Together they approached the W. E. Upjohn Institute with an opportunity we could not refuse. They arranged partial funding for the state study, monitored the project, and contributed much of their own time to aid the effort. We also thank the members of the MOICC Statutory Committee and the MOICC Technical Steering Committee for the review functions they provided for the state report. We have borrowed freely from that earlier work in completing this monograph. Without the encouragement and support of MOICC this study would never have been attempted.

Numerous people shared their technical expertise with us. In particular, we acknowledge the critical input of William R. Tanner, Productivity Systems, Inc. He served as a consultant and helped us bridge the gap between the technical aspects of robotics and their likely economic impacts. Significant contributions were also made by the staff of Robotics International, a division of the Society of Manufacturing Engineers. Various members of the staff of the Office of Economic Growth and Employment Projections, U.S. Bureau of Labor Statistics, provided timely assistance—both with access to data and their expert knowledge of the Occupational Employment Survey. Abel Feinstein and other members of the research staff of the

Michigan Employment Security Commission were also very helpful. Of course, the most significant input to this study remains the various robot manufacturers, robot users, and other robotics experts who so graciously agreed to interviews and who wish to remain anonymous.

Appreciation is also extended to our research colleagues at the W. E. Upjohn Institute. They served as sounding boards throughout our deliberations, contributed their own ideas, and clarified a number of key points through informal discussions. We also thank the loyal support staff of the Institute, especially our research secretaries, Irene Krabill and Denise Duquette. They both assisted with the research and the manuscript wherever and whenever needed. Denise also prepared drafts of many of the bibliographic annotations.

Last we express our gratitude to the Director of the W. E. Upjohn Institute, the late Dr. E. Earl Wright. His counsel supported us, his leadership inspired us, and his confidence in our abilities enhanced our efforts. We miss him deeply but will try our utmost to carry out the kind of practical and meaningful policy research that Earl supported. We hope he would have been pleased with this effort.

H. Allan Hunt
Timothy L. Hunt

Executive Summary

The human resource implications of the so-called robotics "revolution" are explored in this monograph. Specifically, we estimate the job creation and job displacement potential of industrial robots in the U.S. and Michigan by 1990. The study is targeted for policymakers and social researchers, particularly those involved in employment and training questions associated with robotics.

Given the intense media hyperbole and the lack of hard information about robots, it is necessary to develop a broader, more objective perspective of the coming changes before proceeding. First, we submit that robots are simply one more piece of automated industrial equipment, part of the long history of the automation of production. We also argue that the introduction of any new manufacturing process technology is evolutionary rather than revolutionary. There are physical, financial and human constraints on the rate of change in process technology as it is actually applied.

Second, there appears to be a significant lack of understanding that one of the consequences of a growing, dynamic economy, one that makes more goods and services available to all of us through the productivity gains of its workers, is job displacement or the elimination of some jobs through technological change. Concomitantly, we know that other jobs are being created, sometimes in the very same firms that adopt new technologies and sometimes in altogether new sectors of the economy.

In view of the level of interest in robots, it is surprising that so few industries are actually using robots today and that the proven industrial applications are so limited. Virtually all robots can be found in manufacturing firms, with the primary user being the auto industry. The proven industrial applications of robots are welding, painting, and various pick-and-place operations, while assembly tasks hold promise for the future.

We estimate that sales of robots by U.S. producers in 1982 approximated or slightly exceeded the 1981 sales level of $150 million or 2,100 units. By the end of 1982 that implies a total of 6,800 to 7,000 robots were operating in U.S. factories. We also estimate that employment in U.S. robot manufacturing today is roughly 2,000 workers nationwide. This should make it clear that most of the employment impacts of robotics are in the future.

We expect strong growth in the utilization of industrial robots in the decade of the 1980s. By 1990 the total robot population in the U.S. will range from a minimum of 50,000 to a maximum of 100,000 units. Given our estimate of the year-end 1982 population of about 7,000 units, that implies an average annual growth rate of between 30 and 40 percent for the eight years of the forecast period, or roughly a seven- to fourteenfold increase in the total population of robots.

In terms of gross displacement (the elimination of job tasks rather than actual layoffs of workers) we estimate that robots in the U.S. will eliminate between 100,000 and 200,000 jobs by 1990, with roughly one-fourth of that total in the auto industry. In relative terms, job displacement rates due to robots will not be a general problem before 1990 in the U.S., although there will be particular areas that will be significantly affected. Chief among these is the auto industry where from 6 to 11 percent of all operatives and laborers will be displaced by 1990. The results are particularly striking in occupations such as painting and welding for which today's robots are so well adapted. We project that 15 to 20 percent of auto welders and 27 to 37 percent of auto painters jobs will be displaced. Geographically, states such as Michigan, especially the southeastern quadrant with its heavy dependence on autos, will suffer greater displacement than other states or regions.

We do not believe that this job displacement will lead to significant job loss among the currently employed, however. Even in the auto industry, voluntary turnover rates historically have been sufficient to handle the reduction in force that might be required, and the new GM-UAW agreement appears to provide adequate job security measures. However, new labor market entrants may find more and more factory gates closed. Therefore, if there is an increase in unemployment as a result of the spread of robotics technology, we fear the burden will fall on the less experienced, less well educated part of our labor force.

In terms of job creation, we foresee the direct creation of about 32,000 to 64,000 jobs in the U.S. by 1990 in four broad areas: robot manufacturing, direct suppliers to robot manufacturers, robot systems engineering, and corporate users. The largest single occupational group of jobs created by robotics will be robotics technicians—those persons with the training or experience to test, program, install, troubleshoot, or maintain industrial robots. The next most important occupational group is graduate engineers. These will be mostly electrical and mechanical engineers. Together, engineers and technicians account for over one-half of the jobs created.

We anticipate that most robotics technicians will be trained in community college programs of two years duration. The exception is in the auto industry where this requirement will continue to be met through retraining existing members of the UAW Skilled Trades Council without substantial outside hiring. The extent to which other industries will follow a retraining strategy is unknown today. There does not appear to be a supply problem for robotics technicians, as the community college system gives every indication that they will be ready and willing to train whatever numbers are needed. In fact, our current concern is that they may, in some instances, be increasing the supply too rapidly.

The supply of engineers may be more of a problem because there is already a clear shortage of engineers nationwide, so we start from a deficit position. In addition, we face the challenge of other likely engineering demand increases as well as the historical instability of engineering enrollments. Thus it is quite likely that a shortage of engineers could compromise the expansion of robotics technology.

The most remarkable thing about the job displacement and job creation impacts of industrial robots is the skill-twist that emerges so clearly when the jobs eliminated are compared to the jobs created. The jobs eliminated are semiskilled or unskilled, while the jobs created require significant technical background. We submit that this is the true meaning of the so-called robotics revolution.

Contents

Contents

1
"The Robots are Coming"

Introduction

In the past year or so there have been cover stories or special reports about robots in *Time, Newsweek, Fortune, Business Week,* and *The Wall Street Journal,* among others. Indeed, the existence of a robot "revolution" in our factories appears to be treated as a fact in the popular media. Yet there is surprisingly little information available about the possible social and economic implications of robots. How many robots are toiling in our factories today? Which jobs and how many will be done by robots that were once done by human workers? What new jobs and how many will be created by robots? In such an information vacuum it is easy to exaggerate or to misunderstand the few facts that are available; possible even to inadvertently mislead policymakers and the general public as to the impact of robots.

A recent study by Pat Choate warns of the imminent robotization of American factories. He says "the speed and force of this change will be awesome." (Choate, p. 13) He concludes, "As the economy robotizes and domestic jobs are lost to foreign production, 10 million to 15 million manufacturing workers and a similar number of service workers likely will be displaced from their existing jobs. Much of this displacement will occur in the mid- to late 1980s." (Choate, p. 2) Yet nowhere in the study does Choate really say how many jobs will be specifically lost to robots.

On the other hand, Cetron and O'Toole, in their publications on the jobs of tomorrow, predict that millions of new jobs will be created by these same robots. According to them, "there will be as many as 1.5 million robotics technicians on the job in the U.S. alone by 1990. . . ." (Cetron and O'Toole, 1982a, p. 12 and 1982b, p. 259) These technicians will be needed for maintenance of robots for the most part. In a recent issue of *Newsweek,* which highlighted the growth industries and jobs of the future, the work of Cetron and O'Toole and others was referenced. That article included an estimate of total employment in industrial robot production in 1990 of 800,000. ("Growth Industries of the Future," p. 83) If these numbers are believable, then over 2 million U.S. workers will be building or maintaining robots by 1990. At the same time, millions of other workers could be displaced by those robots.

Policymakers, lacking adequate information, must make do with whatever is available. Under these circumstances, even the Secretary of Labor can be misled. In a speech to the Productivity Advisory Committee, Secretary Donovan said, ". . .there will be a major shift from production-line workers to versatile workers able to program, repair, and service the array of robots on the factory floor. In fact, by 1990, half of the workers *in any factory* may well be engineers and technicians and other white collar specialists, rather than the current blue collar workers." (emphasis added)

This small sampling of currently available hyperbole about industrial robots contrasts sharply with the facts, in our judgment. The Robot Institute of America, the industry trade association of robot manufacturers and users of robots, predicts that there will be 75,000 to 100,000 robots in U.S. factories by 1990. (Robot Institute of America, p. 30) Indeed, even the most optimistic robot industry experts foresee no more than 150,000 robots by 1990. In interviews that we conducted, robot manufacturers were certainly enthusiastic about the growth prospects for their industry, but they deplored the "off-the-wall" predictions appearing in the popular media.

In any case, the application of as many as 150,000 industrial robots will not support cataclysmic employment impacts, either in terms of job creation or job displacement. It is not reasonable to think that 1.5 million technicians are needed to maintain 150,000 robots, nor is it reasonable to suppose that 150,000 robots will displace millions of workers. Perhaps it makes interesting reading to claim that by 1990 employment in robot manufacturing will approximate 800,000 people. But such a figure would surpass current U.S. employment in the motor vehicle industry. Even more startling, a figure of 1.5 million robotics technicians by 1990 would surpass current U.S. employment of *all* engineering and science technicians. While these and other wild claims about the impacts of robots may attract considerable media attention, they do not square with the facts, as we shall demonstrate in this monograph.

We agree that the robots are coming, but the near term employment impacts will not be overwhelming by any means. The impact of robots will be felt gradually and cumulatively through the years, an evolutionary rather than a revolutionary process. While these statements may not make headlines, we believe they can be shown to be accurate. In our opinion, the recent intense media attention on robotics may have seriously confused the issues and the policymakers.

Scope and Purpose of the Study

This monograph will explore one aspect of the evolution of technology, the application of industrial robots to the manufacturing process. We focus on the human resource implications of the industrial utilization of robotics technology rather than on the technology itself. More specifically, we estimate the job creation and job displacement potential of industrial robots in the U.S. by 1990. We also derive estimates of the impacts of robotics on one state in the nation, the State of Michigan.

Robotics technology is important to Michigan for at least two major reasons. First, Michigan has traditionally relied on the "metalbending" business for a large share of its manufacturing exports. In particular, the dependence of the Michigan economy on auto and auto-related manufacturing is well-documented. This focus has led to a major concentration on manufacturing process technology as well. Thus Michigan already has a very substantial commitment to manufacturing and to manufacturing process technology.

Second, in 1981, Governor Milliken designated robotics technology as the highest priority in the drive to rebuild the Michigan economy with a high technology base. (Milliken, 1981a, pp. 14-15; Milliken, 1981b, p. 13) Of course, the established stake in manufacturing process technology had a role in the selection. So did the circumstance that the auto industry, upon which Michigan has depended for so long, is the leader in the application of industrial robots to the manufacturing process. It was fairly obvious that industrial robots constituted a threat to the Michigan employment base. It was also obvious that the domestic auto industry had been facing intense competitive pressure from the Japanese, and that part of the Japanese cost advantage was emanating from their superior productivity. This in turn could be attributed to the Japanese use of industrial robots, among other factors.

In the face of this situation, the Governor's High Technology Task Force elected to try to make Michigan a world class center of excellence in manufacturing process technology, including but not limited to robotics technology. The centerpiece of this effort has become the development of the Industrial Technology Institute as an independent non-profit corporation designed (1) to foster basic and applied research in manufacturing process technology, including the social and economic implications thereof, and (2) to provide practical assistance to Michigan manufacturers in both adopting and producing new manufacturing process technology. (Industrial Technology Institute, p. ii)

Because of the various initiatives of the State of Michigan and the belief that robotics technology might significantly affect the state's economy, the Michigan Occupational Information Coordinating Committee (MOICC) asked the W. E. Upjohn Institute to look at the labor market implications of robotics in order to provide a base upon which human resource planning could proceed. The present monograph contains much of the information reported to MOICC in the Michigan study, but the major focus is on the national estimates. Thus, we regard the present volume as an extension of the earlier work.

This study is specifically targeted for policymakers and social researchers, particularly those involved in employment and training questions associated with robotics. No prior knowledge of industrial robots is assumed or needed. Technical questions about industrial robots are discussed only to the extent necessary.

There are precious little hard data about industrial robots today. Our data were gathered through published sources and through interviews with robot manufacturers, corporate users of robots, and other experts. While some judgment was undeniably necessary, we attempted to maintain objectivity throughout our efforts. Our methodology and judgments are explicitly stated in the study. This reflects our hope that this study will lead to other efforts to improve the understanding of the social and economic impacts of industrial robots.

A consistent framework is utilized in the study to evaluate the social and economic implications of industrial robots, particularly the job creation and job displacement caused by industrial robots. That means, for instance, that our projections of the population of robots in 1990 are consistent with our estimates of job displacement and job creation in that same year. Actually, we provide a range for the estimates because of the uncertainties involved, but the point is that the projections are consistent and comparable. This is very helpful in avoiding unrealistic or exaggerated conclusions.

The outline of the study is as follows. In chapter 2 we present a selective review of other forecasts and then our forecast of the U.S. robot population in 1990. The chapter concludes with the derivation of the 1990 projected Michigan robot population. In chapter 3 we discuss the jobs to be eliminated by the robot population projected in chapter 2. That includes not only the number of jobs involved but also the specific occupations. In addition to this examination of job displacement, there is also a discussion of the possible unemployment impacts of robots. Chapter 4 is organized similarly but discusses the jobs that will be created as a result of industrial robots. In both chapters, the focus is on the United States and the State of Michigan. The conclusions of the study are presented in chapter 5.

Given the current lack of information about industrial robots, an annotated bibliography is also provided as part of the study. It is not necessarily complete, nor does it include the popular news magazines or many of the technical journals. However, it is, to the best of our knowledge, the first compilation of an annotated research bibliography on the social and economic impacts of industrial robots. We hope the interested reader can use the annotations to identify items of interest; they cover a broad range, from the highly technical and mathematical economic literature of technological change to simple descriptions of robot characteristics.

In this introduction, the basic facts of robots are discussed first: What is a robot? What work can a robot do? Where are they currently being used? Then the place of robots in production technology is assessed. Since robots are new technology, we discuss the development of two other related technologies, digital computers and numerically controlled machine tools. Next some historical antecedents, including the automation scare of the early 1960s, are considered. These suggested analogies will hopefully lead to some common ground upon which to develop a more dispassionate view of today's new technology—industrial robots. Finally, we conclude chapter 1 with a discussion of a major study which has examined the job displacement effects of robots in great detail: the Carnegie-Mellon study. We believe misinterpretation of that study is responsible for some of the misunderstanding about industrial robots in the popular media.

What is a Robot?

Complete data on current installations of robots in the U.S. are not available. In part, that can be accounted for by confusion in defining exactly what constitutes a robot. A very broad definition originated with the Japan Industrial Robot Association, while the narrower definition used throughout this study originated with the Robot Institute of America (RIA) in 1979. The RIA definition was adopted by the 11th International Symposium of Industrial Robots held in Tokyo, Japan in October 1981. However, it should be understood that international comparisons are still treacherous, and RIA and others have had to reevaluate the U.S. robot population. There is still not total agreement about U.S. installations of industrial robots and no one can be certain exactly how many robots there are in the U.S. today.

The official RIA definition, now accepted internationally, is as follows:

> A robot is a reprogrammable multifunctional manipulator designed to move material, parts, tools, or other specialized devices through variable programmed motions for the performance of a variety of tasks. (Robot Institute of America, p. 1)

The key to this definition is that a robot is a reprogrammable, multifunctional manipulator. A robot can perform the same task on identical workpieces repetitively; it can perform different tasks on the same workpiece; or it can be reprogrammed to perform entirely new tasks.

Unlike R2D2 and C3PO of the movie *Star Wars,* however, robots of today are essentially "dumb machines." They are generally immobile, they usually lack any visual or tactile sensory perception, and they cannot adapt to their environment in any way whatsoever. Generally they are no faster than human workers, but they are tireless. In layman's terms, that means a robot can reproduce a specific range of motions for which it has been programmed, but it does not know if it is really holding the part it is supposed to be or if the work was done correctly. Because of the robot's limitations, it must be carefully interfaced with other equipment using mechanical and/or electrical switches to prevent disasters, and procedures must be established to verify the performance of the robot.

Essentially, then, robots are stationary machines with a manipulator arm that can perform motions repetitively and tirelessly. Unless the workpiece arrives at the exact location for which the arm is programmed, however, the robot will fail. If the workpiece is not of the size expected, or is oriented in the wrong position, the robot will fail. The bottom line is that today's robot can only operate in a carefully structured and oriented world. Furthermore, although the literature makes much of the reprogrammability of robots, relatively few robots today are truly reprogrammed. Minor alterations may be made in the path of the manipulator of a welding robot, but most of today's robots perform the same program over and over and over again.

RIA's 1981 survey reports 4,700 robots in the U.S. by functional application area. (Robot Institute of America, p. 3) By the end of 1982 we estimate that 6,800 to 7,000 robots were operating in U.S. factories. This should make it clear that most of the employment impacts to be discussed are in the future. The growth in application of industrial robots and the implications of that growth both have to be projected because of the very limited empirical base to date.

Robots perform a great variety of tasks today, but most are simple pick-and-place maneuvers such as loading or unloading machines, palletizing, etc. A common sequence might be as follows: the robot picks up the workpiece at a predetermined location, reorients it, places it in a machine tool for processing, removes it after processing, reorients it once again, places the item at a second predetermined location and returns to the beginning. There are also sophisticated welding robots in which the manipulator (arm) can be programmed to follow a continuous path through space instead of simply going to various predetermined points. Control of the entire path of the arm also facilitates spray painting or application of other finishes.

In the auto industry, welding applications of robotics dominate today because auto production is particularly amenable to spot welding robots. There are only a limited variety of auto bodies, the assembly line can pre-position the parts precisely, and the environment can be perfectly organized because the nature of the work does not change. In short, it is a dull, repetitive, hazardous task that is ideally suited to today's robots. For these reasons, automakers are robotizing assembly line welding operations as normal retooling is done.

There are also pilot applications of robots for assembly tasks. However, assembly is generally a very complex task for today's "dumb" robots that cannot tell when the task is done correctly and must operate in a perfectly oriented and organized environment. Suffice it to say here that assembly robots are viewed as the number one growth application of the future. There are considerable ongoing research and development efforts in this area, but presently robots cannot perform most assembly tasks with consistency in an industrial environment at a reasonable cost. The trade literature implies that all of the problems will be solved very soon, and assembly robots will shortly thereafter proliferate in factories all over the world. Others are not so certain.

In sum, the proven applications of robots today are welding, painting, and various pick-and-place operations, while assembly tasks hold promise for the future. Given all of the media attention to robots, it is surprising that there are so few actually in operation. Part of the reason is to be found in the limited industrial applications perfected so far. For a more thorough technical (yet accessible) discussion of robot applications and capabilities, the interested reader should consult the book listed in the bibliography by Joseph L. Engelberger, generally acknowledged as the father of robotics.

Robots in the Productive Process

The auto industry is the primary user of robots today. In fact, the auto industry pioneered many of the current robot applications and continues considerable research and development efforts in the industrial application of robots. Virtually all robots today are utilized in manufacturing firms, and the bulk are located in what might loosely be called metalcutting or metalbending industries (sometimes referred to as the metalworking sector—fabricated metal products, machinery, transportation equipment) and, to a lesser extent, in instruments and related products. Again, the surprise is that so few industries are actually using robots, but it is also true that these industries are particularly concentrated in the five Great Lakes States.

Robots should be viewed as another form of automated equipment. Generally, we can think of two extremes: custom production or dedicated automation. In custom production, general purpose machines are usually hand operated by skilled workers to produce a single item or small lots of that item. Capital equipment costs may be low but total unit costs are high because set-up time can be considerable, individual machining can be a demanding and time-consuming task, and all of the costs must be spread over a very small number of units produced. At the other extreme stands dedicated (or hard) automation, where the initial fixed capital investment can be quite high but total unit costs are typically very low because the automation of production increases speed and insures constant quality. The highly specialized equipment (dedicated automation) is set up once and thereafter production of a single product can flow continuously.

Robots are not identified with either of these extremes. Set-up time for a robot exceeds that of a human operator in custom production, and the speed of a robot is no match for dedicated automated equipment. Instead, robots are a compromise between these two extremes in terms of cost, flexibility and capability. The fixed capital costs of a robot installation exceed that for custom production but are less than dedicated automation; total unit costs are likewise between the two extremes. In terms of *capability,* robots are no match for the subtle skills of a precision machinist, nor can a robot repeat a single task as perfectly as highly specialized automated equipment.

In terms of *flexibility,* the robot once again is no match for a skilled human operator that can adjust a workpiece, correct a minor flaw, and carefully check each and every piece as it is produced. On the other hand, the robot can do different tasks (if it is preprogrammed for those tasks), unlike dedicated automation which is capable of producing a single product only. Specialized hard automation sometimes must be scrapped when the product is changed, whereas in theory the robot can be reprogrammed to perform a new task at any time.

Despite the fact that robots represent a compromise between the extremes of custom production and dedicated automation in terms of cost, capability and flexibility, robots today are being applied primarily in mass production facilities where the human worker or the type of work itself already limits the speed of the overall facility. Thus they are serving primarily as a less expensive alternative to dedicated automation rather than being applied to automate batch production facilities. The robot, once installed, appears to be just an extension of the dedicated automation.

Frequently, one robot that operates alone in the sense that it is not interfaced with other robots but only with the plant equipment which it services is termed a stand-alone unit or robot. In this lexicon, a robot system, then, is simply two or more robots that are integrated with each other and the plant equipment as necessary. Neither stand-alone robots nor robot systems require central computer control over the entire operation, although sufficient limit switches are needed. Stand-alone robot installations dominate today and will continue to do so, at least through the mid-1980s; but robot systems will likely become more important later in the decade.

Some experts think that the greatest potential for robots in the future is the automation of small batch production facilities. (Ayres and Miller, 1981-82, p. 42) This encompasses the ability to reduce batch sizes in production that now require mass production or very large batch facilities (i.e., dedicated automation). The concept appears to promise a capability of production of a family of parts or products as the need arises.[1] Such systems are usually called flexible manufacturing systems, but there is no universally accepted definition. It is unclear how the dedicated machinery for fabrication of manufactured articles would be designed for these new systems, but computer control appears paramount because the automation would require off-line programming of robots and possibly other plant equipment to switch from batch to batch. Ultimately, the individual flexible manufacturing systems would be linked together and lead to the completely automated factory, what some people apparently mean by the term "factory of the future."[2]

However, flexible manufacturing systems will not dominate immediately and the completely automated factory is even farther in the future. Bela Gold, an economist at Case Western Reserve who has studied technological change for over 25 years, stresses the many human and economic prob-

1. The forerunners of these systems are machining centers in which one or more robots service various numerically controlled machine tools to produce precision-cut metal parts. Such machining centers are available today.

2. The terms factory of the future, flexible manufacturing systems and others are encountered frequently in the popular media and trade literature, but they have no consensus definitions at this point.

lems in moving toward the factory of the future. (Gold, 1981a, pp. 30-32, pp. 37-38; and Gold, 1979, pp. 298-302, 310-314) But there are also numerous technical problems. Computer memory systems today are quickly exhausted in controlling even a small manufacturing cell, let alone an entire factory. (Albus, pp. 65-67; Alexander, p. 145; and Wisnosky, p. 22) The integration of individual automated systems in factories involves very complex problems of coordination and transfer. Finally, among the technical problems in robots we note that there are no universal grippers, and off-line programming has not yet been perfected. (Gevarter, p. 37) Today's continuous path robots, for the most part, are "taught" their work task by physically moving the manipulator through the desired sequence of motions.

Our study is focused on the development and introduction of industrial robots and robot systems in manufacturing industries by 1990. Flexible manufacturing systems, the factory of the future, etc., are beyond the scope of the study because their impacts lie beyond 1990, except on an experimental basis. We simply do not find that this technology is sufficiently close to routine implementation to make accurate predictions of its extent or its impact at this time.

Technological Analogies

Since the robot industry is very young today but does have a bright future, it is useful to compare it to other analagous technologies. Such analogies do not prove anything, but they can provide a perspective with which to assess the likely development and diffusion of industrial robots. We briefly review the development of digital computers, certainly one of the most significant technologies of several decades, and numerically controlled machine tools, the most closely related capital equipment to industrial robots.

Before beginning, an important distinction is needed between product technology and process technology. As the names imply, *product* technology is the specific technology that is embedded in a final product, such as calculators or TV's, whereas *process* technology is the technology that is embedded in the capital equipment that makes the final products. Robots are definitely process technology and will likely remain so in the foreseeable future. We do not see an early development of an extensive home market for robots. This distinction is important because there is ample evidence that new product technology tends to diffuse more rapidly than new process technology. (Gold, 1979, pp. 183-184; Mansfield, 1971b, pp. 77 and 84; and Sahal, p. 312)

The growth of digital computers from 1961 to 1979 is presented in table 1-1. The year 1961 was selected because that was the first year in which shipments of computers exceeded 2,000 units, roughly the position in which the robot industry finds itself today. The annual percentage increase in the total population of digital computers averaged 26 percent throughout the 19-year period. There were only three years in which annual shipments declined from the prior year level: 1965, 1967, and 1975. Not surprisingly, relative growth was slightly higher in the earlier years when the total population of computers was smaller, but even in the most recent 10-year period, 1969-1979, the annual growth in the population of computers approximated 24 percent.

What does the growth of computers suggest for the growth of industrial robots, if anything? Digital computers can be classified as process technology in that the computer is not a direct part of the final product (microcomputers for the home market are excluded from the data). Rather, the computer provides information processing—cost accounting, recordkeeping, etc.—that in turn supports the production of a final product. The revelation is that computers, widely heralded as the most significant technological innovation of the 1960s and 1970s, expanded at a growth rate of about 25 percent. Yet some are implying vastly higher growth rates for industrial robots.

Table 1-1
Growth in Digital Computers in the U.S., 1961-1979

Year	Annual shipments (thousands)	Total digital computers (thousands)	Percentage increase in total population
1961	2.2	7.6	-
1962	2.3	9.9	30.3
1963	3.0	12.9	30.3
1964	5.3	18.2	41.1
1965	5.0	23.2	27.5
1966	7.9	31.1	34.1
1967	5.9	37.0	19.0
1968	9.5	46.5	25.7
1969	10.3	56.8	22.2
1970	11.5	68.3	20.2
1971	14.9	83.2	21.8
1972	20.8	104.0	25.0
1973	29.3	133.3	28.2
1974	37.9	171.2	28.4
1975	37.4	208.6	21.8
1976	45.0	253.6	21.6
1977	68.7	322.3	27.1
1978	82.1	404.4	25.5
1979	87.0	491.4	21.5

SOURCE: U.S. Department of Labor, Bureau of Labor Statistics, *Productivity and the Economy: A Chartbook,* Bulletin 2084, October 1981, p. 100.

There are important differences between computers and robots that must be mentioned. It was realized almost from the beginning that computers were widely applicable in both business and government, but robots have only limited applications in the manufacturing sector today. An individual firm can potentially use many more robots than computers; however, robots are directly applied to the firm's production technique. This necessitates careful design, application and integration with the existing production process, while computers are really an adjunct to the production process. There are obviously many differences between computers and robots that make comparisons hazardous, but the fact remains that the growth of the most significant recent innovation in process technology spread or diffused at a rate of about 25 percent annually.

The growth of numerically controlled machine tools is examined because they are more closely related to industrial robots. In fact, robots themselves can be regarded as machine tools. There is also an interesting parallel to robotics technology in the batch production mode. As with robots, numerically controlled machine tools were billed as capable of bringing mass production cost levels to batch production processes because of their great flexibility through reprogramming.

Originally, numerical control meant that the machine tool (lathe, drill press, milling machine, etc.) was controlled by instructions contained on paper tape or cards, while today microprocessor control is becoming more common. The aircraft industry, with research support of the U.S. government, developed numerically controlled machine tools to improve the precision of aircraft parts. This new process technology became available commercially in the mid-1950s; it was widely heralded as applicable in industry anywhere metalcutting was done. By the early 1960s, growth in employment of machine tool operators was thought to be seriously threatened. (Macut, pp. 1-6)

The actual growth of numerically controlled machine tools from 1965 to 1981 is presented in table 1-2. Except for the years 1966-68, the growth of numerically controlled machine tools remained under 20 percent annually. In fact, in 7 of the 16 years in the table, annual shipments declined from prior year levels. The annual growth rate was about 15 percent for the entire period, but averaged only 12 percent for the most recent 10-year period. After 25 years, only 3 to 4 percent of all metalcutting machine tools are numerically controlled. In short, the growth of numerically controlled machine tools has been much less than predicted.

There are many reasons why the growth of numerically controlled machine tools fell far short of expectations, but only three will be mentioned here. First, the applicability of numerical control technology to other industries was significantly overestimated. It appears to have no advantage over conventional machine tooling unless great precision or moderate sized batch production (but less than that needed for justification of dedicated machine tools) is required. (Nabseth and Ray, p. 45; and Mansfield, 1971a, p. 201) Clearly, there must be an opportunity to recover the increased capital investment costs of such technology if it is to be efficient.

Second, there was a significant lack of knowledge about numerical control, and the new technology not only altered the basic production structure but also required the new skill of programming. (Nabseth and Ray, p. 52; and Mansfield, 1971a, p. 201) Thus the human resource limitations were important as well. Third, the price of numerical control ($150,000-$200,000 today for just the hardware) was perceived by many small firms as too high. Many small shops simply do not have the capitalization to afford such investments. Even as recently as 1978, in a survey done of small machine tool firms of 50-100 employees who were nonusers of numerical control but likely candidates for utilization of the technology, it was found that over 72 percent of the surveyed firms had not even *evaluated* numerical control. (Putnam, p. 100)

Table 1-2
Growth of Numerically Controlled Machine Tools in the U.S., 1965-1981

Year	Annual shipments (thousands)	Total NC machine tools (thousands)	Percentage increase in total population
1965	2.1	8.1	-
1966	2.9	11.0	35.8
1967	3.0	14.0	27.3
1968	2.9	16.9	20.7
1969	2.4	19.3	14.2
1970	1.9	21.2	9.8
1971	1.2	22.4	5.7
1972	1.6	24.0	7.1
1973	2.7	26.7	11.3
1974	4.2	30.9	15.7
1975	4.0	34.9	12.9
1976	3.9	38.8	11.2
1977	4.5	43.3	11.6
1978	5.7	49.0	13.2
1979	7.2	56.2	14.7
1980	8.9	65.1	15.8
1981	7.9	73.0	12.1

SOURCE: U.S. Department of Commerce, "Current Industrial Reports, Series MQ-35W, Metalworking Machinery," Annual Summaries, 1965-1980, and Quarterly Summaries, 1981.

Once again, too much can be made of the comparison between numerically controlled machine tools and robots, and there are substantial differences as well as similarities. However, the growth and diffusion of numerical control illustrates the general obstacles to the rapid diffusion of process technology in general.[3]

Historical Analogies

The purpose of the foregoing discussion was to develop a more rational perspective of technological change by briefly looking at two earlier new technologies related to robots, whereas the purpose of this section is to briefly discuss economic change in general. The fear of unemployment and massive displacement caused by labor-saving technology is not new. Such fears began with the dawn of the industrial era in the late 18th century; they continue today with the growth of industrial robots.

For example, the U.S. economy recovered very slowly from the deep 1958-59 recession and then experienced another recession in 1961. The "automation problem" was of urgent national concern, and in 1962 the U.S. Congress passed the Manpower Development and Training Act to address the retraining needs of technologically displaced workers. Then, in 1964, the President appointed a National Commission on Technology, Automation, and Economic Progress to determine the impact of automation and technological change on the U.S. economy.

But the economy was already beginning to recover significantly in 1964, and by the time the Commission rendered its final report in 1966, the economy was near full employment. Historical events ultimately obviated the need for and impact of the Commission; the problem seemed to

3. The interested reader should consult the recent works of Sahal and Gold listed in the bibliography for a review of this literature.

have gone away. To no one's surprise, the Commission's conclusion was that a sluggish economy was the major cause of unemployment rather than automation. (Bowen and Mangum, pp. 3-4)

The recessionary phase of any business cycle is difficult and traumatic for workers, particularly in a state like Michigan with its durable goods-oriented economy. The clear danger is that we may wrongly attribute the short run cyclical problem to other factors, such as automation. Walter Buckingham issued a grim forecast at the time of the 1961 recession: "There are 160,000 unemployed in Detroit who will probably never go back to making automobiles, partly because the industry is past its peak of growth and partly because automation has taken their jobs." (Buckingham, pp. 117-118) Subsequently, however, the auto industry set new employment peaks in the middle of the 1960s, and the auto-dominated Michigan economy boomed once again. (Verway, p. 1) We suffered through another such cycle, although attenuated, with the 1974-75 recession. Yet the auto industry went on to its all-time peak employment in 1978.

The general comparison between the early 1960s and the early 1980s appears compelling in our judgment. History does not and will not repeat itself, but history can provide a more objective perspective within which to judge the current (new) situation. Employment in the auto industry may not recover to its 1978 peak, but employment gains will be significant during the recovery phase of the business cycle.

Automation is not the cause of the U.S. or Michigan's unemployment today any more than it was in the early 1960s. That is not to imply that we should take a "rah rah robots" approach to the coming technological change; however, neither should we adopt a doomsday attitude that attributes most or all unemployment during major recessions to automation. In fact, one might plausibly argue that some of our basic industries suffer more today from a lack of automation and the rational organization of that automation vis-a-vis our European and Japanese competitors than from too much automation.

It is possible to develop a more dispassionate view of technological change, or more specifically, of the introduction of industrial robots. First, let us admit that most technological change throughout American history has been labor-saving, and that means job displacement. By job displacement we mean the elimination of job tasks, not necessarily implying worker unemployment. As will be discussed later, they are not the same thing by any means.

The powerful job displacing effect of technological change is illustrated in table 1-3; it lists hypothetical job displacement in manufacturing in the U.S. and Michigan from 1979 to 1990, assuming a fixed output and a continuation of the slow annual growth in output per worker experienced in the late 1970s of 2.1 percent. (U.S. Department of Labor, 1981c, p. 24) The base year employment for the calculations is 1978. Under the *unrealistic* assumption of constant output, if the annual growth in output per worker of 2.1 percent continues throughout the decade of the 1980s, then cumulative job displacement by 1990 will approximate 4.6 million in the U.S. and 265,000 jobs in manufacturing in Michigan.

Stated in relative terms, 22 percent of all existing jobs in manufacturing could disappear by 1990 as a result of increases in productivity. Of course, worker productivity gains are not solely the result of new labor-saving technologies, but the total effect is the same; gains in productivity, whatever the source, can cause considerable and sometimes dramatic displacement effects on the existing job base if they are examined in isolation.

Table 1-3
Illustrative Displacement Impact of General Productivity Gains, Michigan and U.S. Manufacturing

Year	Cumulative displacement	
	Michigan	U.S.
1979	24,772	430,605
1980	49,023	852,167
1981	72,765	1,264,876
1982	96,009	1,668,919
1983	118,764	2,064,477
1984	141,042	2,451,728
1985	162,852	2,830,847
1986	184,204	3,202,004
1987	205,107	3,565,367
1988	225,571	3,921,099
1989	245,606	4,269,361
1990	265,220	4,610,309

NOTE: The 1978 base year employment figures are 1,179,600 for Michigan and 20,505,000 for the U.S., as found in U.S. Department of Labor, Bureau of Labor Statistics, *Employment and Earnings,* May 1981, pp. 39 and 125.

Second, the dramatic job displacing effects of technological change have not caused massive unemployment in the American economic system because in normal times they have been accompanied by significant economic growth, i.e., output has not been constant. Displaced workers are reemployed in other sectors of the economy, or they may gain new jobs in the same firm if demand increases sufficiently after the introduction of new technology. The heart of the problem appears to be the perception that there is only a constant amount of work to be done, so a machine or robot eliminates not only the job task but also the need for that worker. Historically, this has not generally been true.

Third, the association of technological change and economic growth is not just a coincidence; the two are intertwined and inseparable. That is not to imply that adoption of new technologies necessarily insures economic growth, or that displaced workers will always find new jobs. However, it does mean that we all have a vital stake in productivity gains (i.e., in displacing jobs) because that is what allows the *possibility* of economic growth. The price of a growing, dynamic economy that raises incomes and makes more goods and services available to all of us is job displacement, or the elimination of jobs through technological change.

Fourth, although the long-run impact of technological change has been favorable on the American economy, job displacement in the short run can be traumatic for the workers involved, who usually are concentrated geographically and occupationally. Displaced workers may find it difficult to learn new tasks. Severely impacted regions may not have the resources to cope with those displaced. Job displacement in the short run may require significant public and/or private retraining efforts. Furthermore, the public education system must insure that entry-level workers possess the requisite new skills and not old, obsolete skills.

Finally, we must guard against the temptation to view technological change as revolutionary; the fear that tomorrow we will awaken to the unmanned factory and a world of robots without workers. Technological change tends to be evolutionary, especially in process technology. There are physical, financial, and human constraints on the rate of change of process technology. While no one would dispute that computers have changed our world, this has taken a quarter of a century.

In summary, industrial robots are simply one more piece of automated industrial equipment, part of the long history of automation of production. Robots will displace workers in the same way that technological change has always displaced workers. There is a possibility that this job displacement will be a significant problem, particularly in given occupations, industries, or geographical areas. These questions are examined later in the study. There is also the certainty that robots will create jobs, and that also is examined later in the study. Robots will not guarantee economic growth and we cannot be assured that displaced workers will be reemployed, although there is reason for some optimism historically. In the short run, there will likely be some worker dislocation, and that dislocation may be concentrated geographically. Policy issues raised by these changes will be addressed after their magnitude is determined.

The Carnegie-Mellon Study

We conclude this chapter with a discussion of the only study which has examined the job displacement impacts of robots in great detail, the Carnegie-Mellon study. Actually the Carnegie-Mellon study is not one published document, but several that originated from a project in which Robert Ayres and Steven Miller were the principal investigators. (Ayres and Miller, 1981a)

The fundamental basis of the job displacement estimates of Ayres and Miller is a survey of corporate users of robots (with 16 respondents) that asked them to provide estimates of *potential* job displacement in 32 occupations by today's commercially available robots (Level 1) and tomorrow's robots that would be sensor-based with rudimentary tactile and/or visual perception (Level 2). The occupations were chosen by Ayres and Miller as those most likely to be robotized. The responses were weighted by size of firm (six classes) to obtain a weighted average response. These sampled occupations were then combined with other nonsampled occupations (based on similarity) and job displacement estimates were derived for the metalworking sector and for all manufacturing.

Perhaps Ayres and Miller best summarize their conclusions in a *Technology Review* article:

> Based on these results, we estimate that Level 1 robots could theoretically replace about 1 million operators, and Level 2 robots could theoretically replace 3 million of a current total of 8 million operators. However, this displacement will take at least 20 years. By 2025, it is conceivable that more sophisticated robots will replace almost all operators in manufacturing (about 8 percent of today's workforce), as well as a number of routine nonmanufacturing jobs. (Ayres and Miller, 1982b, p. 42)

According to Ayres and Miller, 4 million manufacturing operative jobs are subject to robotization over the next 20 years or more, and all operatives in manufacturing may be replaced by 2025. The emphasis is clearly on theoretical displacement in the indefinite future rather than actual or probable displacement by some specific date.

We doubt that production techniques, even theoretically, are as homogeneous across manufacturing as Ayres and Miller imply; by industry, by size of firm, or by type of product. But those doubts are minor in the context of theoretical estimation of the unbounded future. As Ayres and Miller themselves point out, their estimates are really only rough guesses to obtain "a feeling of how many people will be involved in 'first order' adjustment processes." (Ayres and Miller, 1981a, p. 100)

Ayres and Miller go on to conclude that their study has highly significant policy implications. They talk of an "institutional failure" in that our public education and training programs reflect obsolete rather than emerging needs. (Ayres and Miller, 1981a, pp. 22-23) They are particularly critical of CETA, vocational schools and government occupational forecasters, none of which in their opinion recognize the future employment needs of society. (Ayres and Miller, 1982a, p. 21) Ayres and Miller conclude, "the transition to the factory of the future is occurring now. . . . If appropriate measures are not taken, the nation will experience unnecessary economic distress and lost opportunities." (Ayres and Miller, 1982b, p. 46)

We do not concur with Ayres and Miller that their estimates of theoretical displacement by occupation at some undefined point in the future are proof that our public institutions today are training their clientele in obsolete skills. Furthermore, Ayres and Miller offer no evidence whatsoever about the emerging occupations, so their criticism in that regard is especially puzzling. In our judgment, if policy responses to the challenges of the future are to be formulated, including the possible effects of robotics technology, then the assessment must proceed based upon the most likely or probable events that are expected to occur within a definite time horizon. That is what we will endeavor to do in the remainder of the study.

2
Forecasts of the Robot Population

Unlike the Carnegie-Mellon study, the projections of occupational displacement in this study are the result of first forecasting the U.S. robot population by industry and application areas within those industries. This approach constrains the displacement estimates to reflect the actual expected sales of robots. In this way, a consistent economic framework is established within which it is possible to estimate not only the population of robots and job displacement but also the job creation resulting therefrom. The job displacement and job creation aspects of the development of robotics are discussed in chapters 3 and 4 respectively.

In this chapter, various other forecasts of the robot population which are inputs to our forecasts are discussed first. Then, the specific methodology of this study to forecast the population of robots is developed. That includes selection of the projection date, robot application areas, user industries, and alternative growth scenarios. Next, our forecast of the U.S. robot population is discussed. Finally, the link of our forecast of the U.S. robot population and the Michigan robot population is established and the resultant estimates presented.

There are quite a few forecasts of the growth in the robot population available today. Some of the more prominent overall forecasts are discussed first. Then, three forecasts that provide more detailed information about application areas and/or user industries are examined: the General Motors corporate forecast, the University of Michigan/Society of Manufacturing Engineering Delphi forecast, hereafter shortened to the UM/SME Delphi forecast, and a forecast of the impact of robots on the U.S. auto industry by William R. Tanner and William F. Adolfson.

General Forecasts of the Robot Population

There are no official U.S. government statistics on the robotics industry. The Robot Institute of America (RIA), the trade association of robot manufacturers and corporate user firms, estimated the U.S. robot population at the end of 1981 to be 4,700 units, approximately 20 percent of the worldwide total. (Robot Institute of America, p. 2) Laura Conigliaro, one of the leading investment analysts of the robotics industry and author of a continuing newsletter about robotics, estimated 1981 unit sales at 2,100 with a dollar value of $150 million. (Conigliaro, June 19, 1981, p. 8) Conigliaro points out that the sales revenue of robot manufacturers includes robots and related items such as controls, software, applications engineering, and sometimes other peripheral hardware systems. The data problems are even more complicated because robot manufacturing may be only a small division of a much larger firm, leading to a lack of accounting uniformity in any robot sales estimates. In fact, Conigliaro stresses that past sales of robots are themselves only estimates, such as her figure for 1980 of 1,450 units with a dollar value of $100 million. (Conigliaro, June 19, 1981, p. 2)

Sales expectations for robots in 1982 were originally quite high for a number of reasons. First, the sales growth rate in terms of revenue was approximately 50 percent in 1981. Second, attendance at the industry trade show, Robotics VI, held in Detroit in March 1982, surpassed even the most optimistic projections. (Jablonowski, pp. 163-178) Third, there had been a flurry of announcements by major firms planning robot production—General Motors, General Electric, IBM, United Technologies, Westinghouse, and Bendix Corporation, to name only a few of the potential entrants. Not surprisingly, the industry also has captured considerable media attention in the last year, which may have fueled public expectations even further.

The media attention notwithstanding, most knowledgeable industry people were not misled. In our interviews, robot manufacturers, robot users, and other robotics experts indicated considerable dismay at the media focus and concern that the industry had caught the public's fancy at the very moment that sales were lagging. As early as March 19, 1982, shortly after the Robotics VI conference, *Iron Age,* a respected trade journal of the metalworking sector, presented an analysis of the robotics industry as one that had indeed been popularized, but one which was short on orders. (Obrzut, pp. 59-83) It is also true that the lack of a significant economic recovery anytime in 1982 and continued weakness in the domestic auto industry surprised most of American industry, including robotics, and may have caused unexpected cancellation of some robot orders, delay in others, and failure to close many prospective sales.

We believe 1982 robot sales were approximately the same as those in 1981, or perhaps slightly higher. If this is correct, then the U.S. robot population at 1982 year-end was about 6,800-7,000 units, utilizing RIA's 1981 base of 4,700 units.[1] Actual 1982 robot sales may appear disappointing to some, but in our judgment, flat sales or modestly rising sales in the face of a longer than expected recession reflects economic

1. The RIA estimate is not universally accepted, but it is representative.

strength. The lesson of 1982 is that robotics, as part of the capital goods sector, cannot expect to be immune from recessions. The vulnerability of the robotics industry to recessions will likely increase as robotics expenditures become a more important (and postponable) part of the capital investment plans of user firms.

Overall forecasts of the growth of the robotics industry usually terminate in 1990. For the convenience of the reader and due to the importance of 1990 in our projections, selected estimates of 1990 sales, average annual growth rates, and the cumulative population of robots in 1990 are presented in table 2-1. They are representative of public sources frequently quoted and respected in the industry.[2] Since there is not universal agreement on the current population of robots, comparison of average annual growth rates may be less meaningful than looking at the expected population of robots.

The available estimates of robot sales and population are roughly similar. Conigliaro forecasts a 1990 market of over $2 billion, 31,350 unit robot sales, and a population of U.S. robots of approximately 122,000. Paul Aron of Daiwa Securities, a leading American expert on the Japanese robotics industry, forecasts a 1990 market in the U.S. of 21,575 units worth about $1.9 billion. (Aron, 1981, p. 60) Aron's 1980-1985-1990 sales figures can be extrapolated to obtain 1990 U.S. robot population of 94,000-95,000. The UM/SME Delphi forecast, details of which are discussed later, foresees a 1990 or 1991 market of approximately 33,333 units which implies a U.S. population of robots of

2. There are other forecasts available, primarily private market studies by such firms as Frost and Sullivan, International Resources Development, Predicasts, and others. We did not have primary access to these documents and did not wish to possibly unfairly characterize them by quoting secondary sources. Suffice it to say that these private market studies tend to be optimistic and project 100,000 or more units installed by 1990.

well over 100,000 in 1990 or 1991.[3] Joseph Engelberger, the father of robotics and president of Unimation, Inc., the nation's leading robot manufacturer, predicts an average annual industry growth rate of 35 percent, with possibly 40,000 unit market sales in 1990. (Engelberger, p. 115) Finally, the RIA, in its own survey, foresees a U.S. robot population of 75,000-100,000 units in 1990. (Robot Institute of America, p. 30)

The overall forecasts of the development of the robotics industry are informative and valuable. However, more specific information is needed to provide occupational and industrial specificity for our study. For that reason the GM corporate forecast, the UM/SME Delphi forecast and the forecast by Tanner and Adolfson are presented.

General Motors Forecast

The GM corporate forecast is presented in table 2-2. General Motors plans to increase the number of robots in use from its 1980 total of 302 to 14,000 in 1990 for an average annual growth rate of 47 percent. As of April 1982, General Motors reported a total of 1,758 robots available (in house or in use). Unfortunately, it is not possible to determine exactly how many are actually in operation, but the goal of 1983 would appear to be well within reach.

Beyond 1983, the GM goals may be more challenging. In a government report about the status of the U.S. auto industry released in late 1981 in which three agencies participated, it is suggested that the length and severity of the slump in the

3. The Delphi estimates are derived from information in the study. Robot sales in 1990-91 are nearly $2 billion, the average price is $30,000 in terms of 1980 dollars, 40 percent of all robot sales are a part of systems, and the robot is 30 percent of the cost of the systems. Thus, the nearly $2 billion in robot sales consists of $.6 billion in stand-alone units, $.4 billion packaged for systems, and $933 million of other systems hardware. The $1 billion in sales of robots only (excluding the other systems hardware) can then be divided by the average price of $30,000 to obtain 33,333 units sales in 1990-91.

auto industry has resulted in a serious erosion of the financial strength of the auto firms. (U.S. Department of Commerce, pp. 1 and 7) Postponement of some modernizing investments for purposes of increasing productivity (such as robots) may be necessary in order to preserve the industry's liquidity. (U.S. Department of Commerce, p. 8)

Table 2-1
Selected Estimates of 1990 Sales, Population and Growth Rates of Robots in the U.S.

Source	Unit sales 1990	Value (billions) (1980 $)	1980-90 annual growth rate (percent)	Cumulative population
Conigliaro[a]	31,350	2.0+	38	122,000
Aron[b]	21,575	1.9	36	94-95,000
UM/SME Delphi[c]	33,333	2.0+	45	150,000
Engelberger[d]	40,000	-	35	150,000
RIA[e]	-	-	35-39	75-100,000

NOTE: The 1980-90 annual growth rate and the cumulative population in 1990 are not necessarily stated directly in all of these studies but can be calculated from data that are provided.

a. Laura Conigliaro, *Robotics Newsletter,* Prudential-Bache Securities Inc., January 15, 1982, p. 7 and June 19, 1981, p. 8.

b. Paul Aron, "Robots Revisited: One Year Later," in *Exploratory Workshop on the Social Impacts of Robotics: Summary and Issues,* Office of Technology Assessment, U.S. Government Printing Office, Washington, DC, July 1981, p. 34.

c. Donald N. Smith and Richard C. Wilson, *Industrial Robots: A Delphi Forecast of Markets and Technology,* Society of Manufacturing Engineers, Dearborn, Michigan, 1982, pp. 47-51, and Donald N. Smith, Peter G. Heytler, and Murry D. Wikol, "Sociological Effects of the Introduction of Robots in U.S. Manufacturing Industry," Industrial Development Division, Institute of Science and Technology, University of Michigan, Ann Arbor, Michigan. Unpublished paper presented at the *CAMPRO '82 Conference on Computer Aided Manufacturing and Productivity,* October 1982, p. 7.

d. Joseph L. Engelberger, *Robotics in Practice,* American Management Association, AMACOM Press, New York, 1980, p. 115.

e. Robot Institute of America, *RIA Worldwide Survey and Directory on Industrial Robots,* Dearborn, Michigan, 1981, p. 30.

Table 2-2
Projected Robot Applications in General Motors

Application	Number of robots in use				
	1980	1983	1985	1988	1990
Welding (Arc and Spot)	138	1,000	1,700	2,500	2,700
Painting	47	300	650	1,200	1,500
Assembly	17	675	1,200	3,200	5,000
Machine Loading	68	200	1,200	2,600	4,000
Parts Transfer	32	125	250	500	800
Total	302	2,300	5,000	10,000	14,000

SOURCE: GM Technical Center, Robotics Display, April 1982.

In 1982 there have been media reports of a slowdown in robot acquisitions at GM and other auto firms due to the lack of financial capital, ("A Robotics Mecca in Michigan? Car Sales Must Rebound First") yet GM must more than double yearly acquisitions from 600-700 to almost 1,500 to meet its 1985 goal. If GM is to meet its robot installation goals, the need for some recovery in the auto industry is apparent. From a slightly different vantage point, near term goals are aided by a major retooling effort that GM committed itself to several years ago, while long term efforts require an increasing share of available financial capital and therefore both a larger management and manpower commitment to robot applications.

Insofar as the details of GM's forecast of their robot population are concerned, GM anticipates a significant and dramatic shift in specific application areas. Welding robots represent almost two-thirds of GM's installations today, while they will be slightly less than one-fifth of the installations in 1990. In contrast, assembly robots, an almost insignificant portion of the total now will grow to over one-third of the total by 1990. The growth in painting and machine loading is more steady. However, new installations of both painting and welding robots will level off well before 1990, while almost one-half of the new installations in that same year will be assembly robots.

There are a number of important implications of the GM plans. First, notice that of the 14,000 robots GM expects to have by 1990, approximately 64 percent will be installed after 1985. This fact alone should emphasize the uncertainties and conditional nature of these plans. Second, early arguments for robots have concentrated on elimination of dirty and dangerous tasks. That argument will carry less weight as robots diffuse to assembly operations and become even more important in machine loading. Third, given that GM expects to install 76 percent of its assembly robots after 1985, it appears that successful application of assembly robots in large numbers awaits technological developments and/or reorganization of the factory floor.

UM/SME Delphi Forecast

The UM/SME Delphi forecast of industrial robots, authored by Donald N. Smith and Richard C. Wilson represents another important contribution to our understanding of robotics. The current UM/SME Delphi forecast reports results of three rounds of questioning on many technical, marketing, and sociological aspects of the development of industrial robots. Over 200 questions were asked in round one, while rounds two and three repeated some questions of round one as well as adding supplemental questions suggested by the experts. The total number of participants ranged from 36 to 60, with as many as 90 percent from corporate user firms.

The Delphi technique itself is an iterative forecasting process in which experts independently input their own forecasts of the future by responding to a consistent series of questions. The objective of the Delphi methodology is to gain consensus through iterative polling. The assumption is that the collective opinion of the group is better than that of any single person. It should be mentioned that the current UM/SME Delphi forecast is an interim report and does not yet meet the usual Delphi requirements for consensus and precision.

One pertinent aspect of the UM/SME Delphi forecast for our study is a ranking of the importance of various robot application areas by industry for 1980, 1985, and 1990. Tables 2-3 and 2-4 summarize these rankings for all industry and for autos. Once again, the growth in importance of assembly applications is clear, particularly in autos.

It is even more interesting to examine the percentage or relative usage of robots by application areas and industry. Since the percentage shares remain more or less stable, table 2-5 presents the results for 1990 only. However, it does include all of the industries specified in the UM/SME Delphi forecast—autos, casting/foundry, heavy manufacturing, light manufacturing, electrical/electronic, and the aerospace industry. Although the UM/SME Delphi forecast defined the robot application areas differently here from in the rankings just discussed, it is apparent that welding and painting are more important in autos than elsewhere, while machine loading, press loading, and drilling, routing, and grinding are slightly less important in autos.

Finally, the UM/SME Delphi estimates of the total relative market shares by industry, i.e., the percent of total robot shipments to each of the industrial sectors, are presented in table 2-6 for all of the years reported in the UM/SME Delphi forecast. The auto industry is expected to remain a stable part of the market with slightly less than one-fourth of all shipments. Light manufacturing is expected to have a somewhat larger share, although the UM/SME

Table 2-3
Delphi Forecast: Rank Importance of Robot Application Areas in All Industry, 1980-1990

Application	1980	1985	1990
Pick-and-Place	1	1	1
Machine Loading	2	1	1
Continuous Path (e.g., paint, weld)	3	3	1
Manufacturing Processing (e.g., drilling)	4	5	5
Assembly	5	4	4
Inspection	6	6	6

SOURCE: Donald N. Smith and Richard C. Wilson, *Industrial Robots: A Delphi Forecast of Markets and Technology,* Society of Manufacturing Engineers, Dearborn, Michigan, 1982, p. 52.

NOTE: Ranked from most frequent (1) to least frequent (6).

Table 2-4
Delphi Forecast: Rank Importance of Robot Application Areas in the Auto Industry, 1980-1990

Application	1980	1985	1990
Pick-and-Place	3	3	4
Machine Loading	2	2	2
Continuous Path (e.g., paint, weld)	1	1	1
Manufacturing Processing (e.g., drilling)	4	5	5
Assembly	4	4	2
Inspection	6	6	6

SOURCE: Donald N. Smith and Richard C. Wilson, *Industrial Robots: A Delphi Forecast of Markets and Technology,* Society of Manufacturing Engineers, Dearborn, Michigan, 1982, p. 53.

NOTE: Ranked from most frequent (1) to least frequent (6).

Table 2-5
Delphi Forecast: Relative Importance of Robot Application Areas by Industry, 1990

Application	Percentage of robots within industry category					
	Automotive	Casting/ foundry	Heavy manu- facturing	Light manu- facturing	Electrical/ electronic	Aerospace
Gas/Metal & Arc Welding	11	6	12	9	9	5
Resistance Welding	17	6	6	11	4	21
Machine Loading	23	33	23	23	27	16
Painting	14	6	12	11	9	11
Press Loading/Unloading	11	22	23	17	13	11
Drilling, Routing, Grinding	11	16	12	11	16	21
Other	11	11	12	17	22	16
Total	100	100	100	100	100	100

SOURCE: Donald N. Smith and Richard C. Wilson, *Industrial Robots: A Delphi Forecast of Markets and Technology,* Society of Manufacturing Engineers, Dearborn, Michigan, 1982, pp. 56-58.

NOTE: Totals may not equal 100 percent due to rounding.

Table 2-6
Delphi Forecast: Percent of Total Robot Shipments by Industry

Industry	1979	1980	1981	1982	1983	1984	1985
Automotive	17.8	20.0	22.2	23.3	23.3	23.3	22.5
Casting/Foundry	21.3	19.4	20.0	20.0	14.0	13.3	11.3
Heavy Manufacturing	9.9	9.7	8.9	8.3	8.1	7.5	6.3
Light Manufacturing	36.6	33.3	33.3	33.3	27.9	31.7	25.0
Electrical/Electronic	11.1	11.1	9.8	11.7	9.3	10.0	8.1
Aerospace	0.9	1.1	1.3	1.7	2.1	2.1	2.0
Other	2.4	5.4	4.5	1.7	15.3	12.1	24.8
All Industry	100.0	100.0	100.0	100.0	100.0	100.0	100.0

SOURCE: Donald N. Smith and Richard C. Wilson, *Industrial Robots: A Delphi Forecast of Markets and Technology,* Society of Manufacturing Engineers, Dearborn, Michigan, 1982, p. 51.

Delphi forecast does not provide a specific definition for this industry. Notice also the small market shares expected for aerospace and the electrical/electronic industries.

Tanner and Adolfson Forecast

William R. Tanner, a robotics expert and engineering consultant, and William F. Adolfson have conducted a study of the application of robots in the North American motor vehicle industry for the U.S. Department of Transportation. The report presents a wealth of information about robots in the auto industry not obtainable from any other source.

The estimates by Tanner and Adolfson of the North American robot population in autos for various years are presented in table 2-7. These projections are classified according to various assumptions about conditions in the auto industry and the nation. The "minimum effort" estimates assume continuation of the status quo which includes lagging auto sales and strong foreign competition, at least through the mid-1980s. The "moderate effort" estimates assume a modest recovery in the domestic auto market and some decline in interest rates. The "strong effort" estimates include the moderate effort assumptions and add assumed improvements in general investment incentives such as tax credits and accelerated depreciation allowances. They also anticipate advances in robotics technology which might include low-cost sensory feedback systems. Finally, the "maximum effort" estimates assume, in addition to the foregoing, direct investment incentives for robots and government provision of retraining/relocation assistance for displaced workers. In sum, Tanner and Adolfson forecast a 1990 robot population in the North American auto industry ranging from a low of 18,500 units to a high of 35,700 units.

Tanner and Adolfson also present representative cost breakdowns for a single robot installation for machine loading and for a major robot welding system. These detailed estimates are presented in tables 2-8 and 2-9, respectively. The single installation carries a price tag of $125,000, while the major system of 12 robots costs $2.5 million. The specific cost estimates are not as important as the fact that even in the case of a single robot installation (frequently called a stand-alone robot), the robot itself represents less than one-half of the total cost of the installation. That percentage shrinks to one-third or less in the case of a major robot system. Tanner and Adolfson project the auto industry may have a cumulative investment in robots of $2.3 to $4.0 billion by 1990.

Table 2-7
Tanner and Adolfson: Projected Industrial Robot Population in North American Automobile Industry, 1980-1990

Assumption	1980	1983	1985	1988	1990
Minimum effort	1,065	2,600	4,700	10,800	18,500
Moderate effort	1,065	4,050	7,500	16,200	22,600
Strong effort	1,065	4,500	10,000	20,000	28,000
Maximum effort	1,065	4,500	11,200	25,000	35,700

SOURCE: William R. Tanner and William F. Adolfson, *Robotics Use in Motor Vehicle Manufacture,* Report to the U.S. Department of Transportation, February 1982, p. 100.

All of the available forecasts were valuable aids in the development of our methodology and forecasts. Not surprisingly, however, none were exactly compatible with our need to project application areas with specific occupational and industrial content. This was especially true in view of the need to apply those estimates to the State of Michigan. Furthermore, it is clear, regardless of the desire of policymakers and others for detailed information about the future of robots, that only general and tentative information is possible today.

Table 2-8
Tanner and Adolfson: Cost Elements of Typical Single Robot Installation for Machine Tending

Element	Representative cost (thousands)	Percent of total systems costs
Robot	55	44
System design	10	8
End-of-arm tooling	5	4
Conveyors and part orienters	15	12
Controls and interfacing	7	6
Safety devices, guard rails, etc.	5	4
Rearrangements and site preparation	5	4
Equipment relocation and revision	5	4
System installation, robot programming and debugging	5	4
Personnel training	3	2
Efficiency and production losses during start-up	10	8
Total	$125	100

SOURCE: William R. Tanner and William F. Adolfson, *Robotics Use in Motor Vehicle Manufacture,* Report to the U.S. Department of Transportation, February 1982, p. 42.

Table 2-9
Tanner and Adolfson: Cost Elements of Typical Major Robot System for Body Assembly Welding

Element	Representative cost (thousands)	Percent of total systems costs
Twelve robots	850	34
System design	250	10
Welding guns, transformers and controls	150	6
Conveyors	150	6
Locating and positioning fixturing	250	10
Controls and interfacing	200	8
Safety devices, guard rails, etc.	50	2
Site preparation	150	6
System assembly, tryout and shipping	250	10
System installation, robot programming and debugging	100	4
Personnel training	25	1
Efficiency and production losses during start-up	75	3
Total	$2,500	100

SOURCE: William R. Tanner and William F. Adolfson, *Robotics Use in Motor Vehicle Manufacture,* Report to the U.S. Department of Transportation, February 1982, p. 41.

Upjohn Institute Forecast

The purpose of this section is to present our forecasts of the 1990 U.S. robot population and the Michigan robot population. That requires selection of a projection date, economic scenarios, robot application areas, and user industries. Second, a logical relationship between the robot population in the U.S. and the State of Michigan is developed to specifically estimate Michigan's robot population.

Methodology

Unquestionably, the easiest methodological decision was the selection of a projection date. Few forecasts of the impacts of robotics have ventured beyond 1990. Post-1990 technology is problematical, and it is difficult enough just to project the impact of an infant industry to that date. In short, the terminal projection date of this study of the human resource implications of robotics is 1990.

Single point estimates of future unit sales of robots, dollar value of sales, and population of robots are inadvisable. Our judgment is that such specificity is misleading, however well-intentioned the estimates may be. Consider, for instance, the impact of a 5 percent variation in Engelberger's expected 35 percent average annual growth rate in the population of robots, 1983-1990, assuming the 1982 year-end stock approximates 6,800. If the growth rate is 30 percent (a healthy growth trend for any industry) the stock of robots at the end of 1990 is 55,470. On the other hand, with a growth rate of 40 percent, the stock of robots at the end of 1990 would be 100,354. A variation of plus or minus 5 percent around Engelberger's expected growth rate of 35 percent causes nearly a 100 percent variation in the 1990 stock of robots. Of course, such a result reflects the small existing stock of robots and the assumption of exponential growth. Nonetheless, this example clearly illustrates the difficulty with point estimates for the population of robots.

Two scenarios are developed in this study. The low-growth scenario for robotics assumes relatively high interest rates and slow real GNP growth, approximating the late 1970s annual average of 2.0 percent. The implications for the auto industry in the low-growth scenario are some recovery of domestic auto sales from their current depressed levels, but failure to achieve the vigorous rebound that has so often characterized auto sales in the past. The high-growth scenario for robotics envisions further interest rate declines and real GNP growth that approximates the post-World War II annual average of 3.5 percent.

Neither scenario includes any specific assumptions of breakthroughs in robotics technology, although clearly, as will be pointed out later, the growth in importance of assembly robots requires some technological improvements. There are three reasons why specific technological assumptions appear unwise. First, the available economic research indicates that there can be a considerable lag between a specific technological breakthrough and successful marketing of the resultant product, particularly in the case of process technology.[4] Second, the same economic research indicates there can be considerable delay in application of new process technology across industries even after successful adoption in one industry. The reason is that the technical requirements of each industry tend to be unique, and cross-industry adoption frequently requires further adaptation of the original process. Third, as stated by the Chairman of the Board of Prab Robots, Inc., "the present level of robot technology seems to be much more than U.S. industry can readily absorb." (*Prab Robots, Inc. Annual Report, 1981,* p. 4)

The implication, strongly confirmed by our interviews, is that diffusion of robotics technology will be limited more by a lack of human understanding of the *existing* technology than by a lack of new hardware. Perhaps surprisingly, this lack of understanding applies even to the major corporate user firms of today, including the auto industry. In any event, it appears unwise and unnecessary to make any specific technological assumptions for the forecast period, except as already noted.

---

4. See the works of Mansfield, Sahal and Gold for an elaboration of these points.

Several other factors specifically included in our alternative growth scenarios must also be mentioned. First, continued strong foreign competition in autos is expected throughout the decade. Second, special investment incentives for robots are unlikely. Third, there is the usual caveat about unforeseen economy-wide shocks that may completely invalidate the forecast.

In short, our low-growth scenario for robotics presumes a continuation of slow GNP growth, lagging auto sales, and high interest rates, while the high-growth scenario maintains a return to our historical GNP growth trend and decline in interest rates to more reasonable levels as well. Obviously, the extremes of major economic depression or "booming" reindustrialization are avoided.

The selection of specific robot application areas to be enumerated in this study is important because the application areas must have related occupational content to be meaningful. The need for occupational content coincident with available data restricts the application areas to five in the study: welding, assembly, painting, machine loading/unloading, and other. Clearly, more specificity would be desirable, but it is necessary to develop the robot data in a way that maximizes the comparability with employment data. Thus it seems preferable to aggregate robot application areas somewhat differently from other authors for our purposes.

The robot application area of welding includes resistance or spot welding and arc welding. Resistance welding applications dominate in industry today, especially in autos. Utilization of arc welding robots will grow in the 1980s now that seam-following arc welding robots are available, although there is still some disagreement about the likely extent of that growth.

Assembly robots exist in research and development laboratories and pilot applications in industry, but most assembly operations are incredibly complex for today's robots. The future importance of assembly robots depends on several intertwined factors: the adequacy of sensory perception, adaptability to the workplace environment, and the rationalization or orderliness of the workplace environment. Rudimentary vision systems are available, but adaptability remains extremely limited and reliability has yet to be conclusively demonstrated in an industrial environment. In part, it is simply a problem of consistency—the robot or robots must assemble a workpiece of perhaps 8 to 16 parts (or more) perfectly for 14 hours a day.

In a joint project, Unimation and General Motors have developed a robot for small parts assembly called the Programmable Universal Machine for Assembly (PUMA). Initial applications of the PUMA robot are now in progress. Engelberger describes these robots as designed to do automotive subassemblies; they will work alongside their human counterparts doing the simpler assembly tasks. (Engelberger, p. 137) He also believes that assembly robots are closer to reality than any other new application. (Engelberger, pp. 134-135)

The robot application area of painting includes robots that are capable of spray painting and application of other finishes, coatings, and sealants. It should be noted that the workplace environment here is particularly unhealthy for human workers. In addition, consistency of the final product can be improved with robot application, and significant savings in materials are also reported. Painting now ranks with welding as a proven robot application.

The application area of machine loading/unloading is very broad in this study. It encompasses casting, forging, press loading, machine tool loading, and heat treatment. Machine loading robots are currently more important, both absolutely and relatively, outside the auto industry, and that relationship is expected to continue.

The final category of "other" includes robots that are used primarily for parts transfer or material handling, inspection and other new application areas. The auto industry does not foresee a large role for robots in parts transfer in their operations, but the possibilities may be significant in other manufacturing areas.

The specific application areas of robots are closely related to the industries that will most likely use robots. Virtually all robots today can be found in the manufacturing sector, and that is not expected to change significantly in the forecast period. In this study, industrial detail is shown for autos and all other manufacturing only. The dichotomy of autos and all other manufacturing was chosen for a number of reasons. First, considerably more information is available about the auto industry. It is not only the largest current user of robots, but also the auto firms have publicly announced their future plans for utilization of robots. Second, the auto industry is dominant in the State of Michigan, and it is only in the auto industry that robots will have a significant impact in the state during the forecast period, as discussed later. Third, since the auto industry is beset with such severe problems and challenges at the present time, it may well serve as a prototype for the general impact of robotics on manufacturing technology in the U.S.

U.S. Robot Population in 1990

Our forecast of the U.S. robot population by application and industry is presented in table 2-10. Although we utilized all available information in formulating these projections, including other forecasts and our interviews with leading experts in the industry, the forecast represents our own judg-

Table 2-10
Forecast of U.S. Robot Population by Application, 1990

Application	Autos		All other manufacturing		Total	
	Range of estimate		Range of estimate		Range of estimate	
	Low	High	Low	High	Low	High
Welding	3,200 (21.3%)	4,100 (16.4%)	5,500 (15.7%)	10,000 (13.3%)	8,700 (17.4%)	14,100 (14.1%)
Assembly	4,200 (28.0%)	8,800 (35.2%)	5,000 (14.3%)	15,000 (20.0%)	9,200 (18.4%)	23,800 (23.8%)
Painting	1,800 (12.0%)	2,500 (10.0%)	3,200 (9.1%)	5,500 (7.3%)	5,000 (10.0%)	8,000 (8.0%)
Machine loading/unloading	5,000 (33.3%)	8,000 (32.0%)	17,500 (50.0%)	34,000 (46.0%)	22,500 (45.0%)	42,000 (42.0%)
Other	800 (5.3%)	1,600 (6.4%)	3,800 (10.9%)	10,500 (14.0%)	4,600 (9.2%)	12,100 (12.1%)
Total	15,000	25,000	35,000	75,000	50,000	100,000

ment. In general, we concentrated on forecasting the individual application areas by industry first, rather than the overall totals. For the convenience of the reader, however, we begin with a discussion of the overall forecast and then proceed to the industries and specific application areas within those industries.

We expect strong growth in the utilization of industrial robots in the decade of the 1980s. By 1990 the total robot population in the U.S. will range from a minimum of 50,000 to a maximum of 100,000 units. Given our estimate of the year-end 1982 population of approximately 6,800-7,000 units, that implies an average growth rate from 30 percent to 40 percent for the eight years of the forecast period, or roughly a seven- to fourteenfold increase in the total population of robots.

It may be worth mentioning that our range for the total population of robots in 1990 is not dependent in any way on the 1982 year-end stock (or some hypothetical growth rate). There is no universal agreement on the U.S. population of robots in 1981, although RIA's estimate of 4,700 units is the one most frequently accepted,[5] and our estimate of 1982 sales may be in error. In short, regardless of near term market conditions and/or re-evaluations of the existing population of robots, we believe our forecast range represents an appropriate and reasonable minimum and maximum for the U.S. population of robots in 1990.

The overall forecast may appear similar to other available forecasts, but it differs from them in at least one major way. Other industry forecasts for the 1990 U.S. population of robots tend to be near 100,000 units or above. Presumably, these are "most likely" or "most probable" forecasts, since

5. There appear to be two problems. RIA's definition of a robot was only officially adopted in 1979. It excluded mechanical transfer devices and thus required downward revisions in the stock estimates. Second, the imports of robots are very difficult to estimate.

they are single point estimates only, whereas our 100,000 unit forecast is a maximum which we are reasonably confident will *not* be exceeded. In other words, we predict strong growth for the robotics industry in the 1980s, but that growth will likely be slightly less rapid than other forecasts would indicate.

There are many factors that support our conclusion of strong growth for robotics, but perhaps less spectacular than generally anticipated. Some of these factors have been discussed previously, but they are mentioned once again to highlight the important points. First, and perhaps most important, American industry lacks trained personnel both to implement robotics technology and to maintain and support that technology once installed. One corporate user reported advertising for a graduate engineer with experience in robotics and then receiving only one application showing any experience whatsoever (and that experience was minor).

Another complaint mentioned in our interviews was that American universities produce engineers who are overly specialized, rather than a generalist who understands manufacturing technology and how to make it work. There were even complaints about the lack of salesmen who truly understand the capabilities and possibilities for utilization of robotics technology. As stated previously, the lack of skilled manpower applies to major current corporate users and to a lesser extent to robot manufacturers. Although educational programs for skilled robotics technicians (two-year degree) are expanding rapidly, the supply of graduate engineers is much less elastic. The lack of engineers with meaningful and practical robotics work experience will likely continue for quite some time. These matters are discussed more fully in chapter 4.

A second factor that will limit the growth of industrial robots is the financial commitment necessary to implement robotics technology. A system of four to six robots can cost in excess of $1 million. Even three stand-alone units can cost from $300,000 to $400,000. Engelberger reports that utilization of less than three robots at a particular location is unwise and uneconomical due to maintenance requirements. (Engelberger, p. 86) Others might place this figure slightly higher, but the important point is that robotics technology requires more than a nominal commitment of funds. Furthermore, although advertised robot prices are falling, the robot itself usually represents less than 40 percent of the total cost of installation, so dramatic price relief is not likely. Finally, our interviews with robot users consistently indicated one warning about the cost of robot installations: be prepared for a longer than expected start-up time. Given that the primary motivation in adopting robots is the labor savings, start-up delays can erode some of the cost savings rather quickly.

The third limit to the growth of industrial robots, closely allied with the financial commitment just discussed, is the management commitment needed to successfully adopt industrial robots. Pilot installations of robots almost invariably identify some part of the factory that can operate in isolation from the rest of the factory to ease the initial introduction of robots (and assure their success), but those types of installations are limited. Eventually, user firms must rethink and fundamentally restructure the factory to accommodate robots. According to Bela Gold, however, the emphasis of American industry on short-run payback does not facilitate such fundamental rethinking. (Gold, 1981a, p. 37)

The UM/SME Delphi survey asked respondents for the payback period required by user industries to justify an investment in a robot. The response was that the required payback period today is two to three years. (Smith and Wilson, p. A-60) More important, the respondents also said that the required payback for a robot investment is expected to remain stable or *decrease* in most industries. (Smith and Wilson, p. A-60) In that light, it should come as no surprise that the Carnegie-Mellon survey found that the bulk of all respondents expected to install robots as retrofits in existing plants over the period 1980-1985. (Ayres and Miller, 1981a, p. 142) The apparent conclusion is that the emphasis of American industry on short-run payback to justify expenditures on new plant and equipment applies to robots as well.

The fourth limit to the growth of industrial robots might be termed general economic conditions. Very few economists expect vigorous GNP growth in the decade of the 1980s, and most would probably argue that even average GNP growth consistent with the post-World War II annual average of 3.5 percent is unlikely. Furthermore, the robotics industry will not likely be immune from the business cycle, so several years of 50 percent growth may be followed by no growth or even sales declines. Although we are aware of the reports that American industry must reindustrialize rapidly to survive in world markets and that such capital investment is inevitable due to the aging of the existing stock of capital in the U.S., we expect economy-wide investment will be more incremental and gradual, consistent with slow GNP growth.

Finally, much more rapid diffusion of robotics technology than earlier process technologies appears highly unlikely. Not only are there significant time lags between innovation and successful marketing but also there can be significant lags between successful adoption in one industry and adoption in other industries, as discussed earlier in the study. More important, and in sharp contrast to the Carnegie-Mellon study, we expect diffusion of robotics technology to be limited primarily to large firms, and perhaps even Fortune 500 firms, for the foreseeable future. Just as small firms have not adopted numerically controlled machine tools, small firms will not risk their very existence by the adoption of robots.

The foregoing limits notwithstanding, we do expect sustained growth for industrial robots. It is only because of the almost euphoric expectations for this industry that we emphasize our doubts: the lack of trained personnel, the large financial and management commitment required, the unlikely prospect of vigorous GNP growth to support robot investments, and the difficulties of diffusion of process technology in general, including the likely prospect that robots will remain large firm "big ticket" items for the foreseeable future. In short, the robots are coming, but we believe the change will be incremental and evolutionary rather than revolutionary.

Turning to the industry forecasts, we project that the 1990 U.S. population of robots in the auto industry will range from 15,000 to 25,000 units. If the auto industry firms were to exactly meet their announced plans, there would be nearly 20,000 robots in the U.S. auto plants by 1990. The range of our forecast thus allows for approximately a 25 percent variation in those plans. It is roughly comparable to the minimum and strong effort forecasts for the auto industry by Tanner reported in table 2-7 (less Canada and Mexico).

This small range, much smaller than for the remainder of manufacturing, reflects our judgment on a number of matters about the auto industry. First, the auto industry is the recognized pioneer and largest current user of robotics technology. Second, the auto industry has undertaken considerable research and development efforts in robotics technology. Third, international competitive pressures and one of the highest average wage rates in U.S. manufacturing lend economic support to the robotization of auto plants. Finally, although considerable retooling of auto plants has already been accomplished to accommodate the new downsized, front-wheel drive, fuel-efficient autos, U.S. auto manufacturers plan strong capital expenditures throughout the decade of the 1980s to continue product changes and meet government-mandated standards. (Arthur Andersen & Co., 1979, p. 14) Of course, if there is no recovery from the current depressed state of the auto industry, then robot investment plans and the very survival of the industry will be jeopardized. However, we do not think such gloomy possibilities are reasonable.

Within the auto industry, the relative magnitudes of the estimates were strongly influenced by the public announcements and plans of the auto firms. Assembly robots are the most important application area within the high range of the estimate, with 8,800 units or 35.2 percent of the total of 25,000 robots in the auto industry, while assembly robots are second in importance within the low range of the estimate. Machine loading/unloading is the most important application in the low-growth case, with 5,000 units or 33.3 percent of the total of 15,000 robots in the auto industry, while machine loading/unloading is second in the high-growth case. Thereafter, the relative rankings are the same in both the low and high range of the estimate, with welding applications third, painting applications fourth and other applications fifth. In the auto industry in the decade of the 1980s, there is little doubt that the proven applications of welding, painting, and to a lesser extent, machine loading/unloading will be pursued most aggressively first, followed by assembly applications later.

The forecasts of the specific application areas within autos and all other manufacturing reflect more technical considerations than anything else. In general, the range of the estimates for each of the application areas in autos is narrower than for each of the application areas in all other manufacturing due to the greater uncertainties in all other manufacturing. The range of the estimates for welding, painting, and machine loading/unloading tend to be narrower than the other application areas in both autos and all other manufacturing because these three application areas are technically feasible today. Likewise, the range of the estimates for assembly robots and "other" robots is broader because assembly robots are not currently proven applications and the other category of robots allows for the development of new applications as well.

The diffusion of industrial robots in all other manufacturing is more difficult to predict than in autos. While the major auto firms have announced their robot investment plans, much less information is available about other industries. Whereas the auto industry is almost totally dominated by large firms likely to adopt robotics technology, other manufacturing is less dominated by large firms. The auto industry is clearly the pioneer in the successful utilization of robotics technology, but its technical applicability to other industries may require further adaptation, and the cost-effectiveness of those applications is not as certain. For these reasons and others, we project a rather broad range for the population of robots in all other manufacturing of 35,000 to 75,000 units in 1990.

Within all other manufacturing, machine loading/unloading applications are expected to continue their dominance in both the high and low range of the estimates with nearly 50 percent of the total population of robots performing machine loading/unloading tasks. Assembly applications are second in the high-growth case, while they are third in the low-growth case. The range of the estimate for assembly robots is especially broad—5,000 to 15,000 units—reflecting both the technological uncertainties and the possible difficulties of adaptation across industries. In that regard, it should be mentioned that research and development in assembly robots is being conducted by the electronics and office computer industries within all other manufacturing.

Welding, a proven application, is the second most frequent application in all other manufacturing within the low range of the estimate, while it falls to fourth in the high range of the estimate as newer applications, especially assembly, become relatively more important. Finally, painting remains in fifth position in both the high and low range of the estimates in all other manufacturing, with 3,200 to 5,500 units expected to be installed by 1990.

As stated earlier, we focused on forecasting the specific application areas by industry rather than the overall totals, so the overall totals by application areas are simply the sum of the individual industry estimates by application areas. Overall in our forecast, it turns out that machine loading/unloading is first, assembly is second, welding is third, and painting and the "other" category exchange the fourth and fifth positions depending on the assumed growth scenario.

Some final comments about our robot forecast are in order. Although autos represent about one-fourth of the current robot market, there is no necessary reason for that relationship to continue. It is reasonable to think that the market share of autos as a proportion of total robot sales will depend on economic conditions in the auto industry itself. Also, there is little reason to select the mid-point of the range of any of our estimates (including the range for each of the specific application areas) as the most likely single point estimate possible. Uncertainties about the development and diffusion of industrial robots are so great that more specificity than the range itself is impossible at this time.

Michigan Robot Population

The forecast of the U.S. robot population in 1990 is used to derive the Michigan robot population in that same year. The specific methodology is illustrated in table 2-11. According to the *1977 Census of Manufactures,* 35.5 percent of all the production workers in the U.S. motor vehicle and equipment industry (SIC 371) are located in Michigan, while only 4.1 percent of production workers in the remainder of manufacturing are located in the state. These percentages, indicating the relative importance of a particular industry in Michigan as a proportion of the same industry in the U.S., are utilized to assign robots to the state by industry and application area.

Table 2-11
Production Worker Employment in Michigan and the U.S., 1977

Industry	U.S. production workers (thousands)	Michigan production workers (thousands)	Percent of U.S. industry in Michigan
Motor vehicles and equipment	727.6	258.4	35.5
All other manufacturing	12,963.4	531.0	4.1
Total manufacturing	13,691.0	789.4	5.8

SOURCE: U.S. Department of Commerce, Bureau of the Census, *1977 Census of Manufactures: General Summary,* Vol. 1, U.S. Government Printing Office, Washington, DC, 1981, pp. 1-59 and 1-94.

Although the foregoing appears to be the only feasible alternative to estimate the number of robots in Michigan in 1990, there are a number of implications and/or limitations that must be explicitly stated. First, this method assumes that the relative importance of autos and all other manufacturing in the state vis-a-vis the nation will remain the same during the forecast period. That is not at all clear. David Verway reports that a centralized auto industry utilizing the Japanese "kan-ban" system of producing and delivering parts exactly when they are needed strongly favors the Midwest, not only because the Midwest and Michigan are already the center of the auto industry, but also because strong import competition on the East and West Coasts argues against expansion there. (Verway, p. 6)

On the other hand, Verway also argues that the passage of domestic content legislation would work against the Midwest because the Japanese would probably locate their U.S. plants near their major markets, the East and West Coasts. (Verway, p. 6) Complicating matters more, GM announced recently that discussions were under way with Toyota for joint production of a subcompact car, probably utilizing two California plants which were only recently shut down. Chrysler recently announced similar joint plans with Mitsubishi for a subcompact automobile utilizing a Missouri plant.

All of these potential locational influences and others cannot be untangled sufficiently to support any other assumption than relative stability in Michigan's proportion of production in the U.S. auto industry during the forecast period. Michigan's proportion of all other manufacturing may fall slightly during the forecast period, but that is of little importance since the number of robots in all other manufacturing in Michigan is expected to be small.

A second implication of the methodology utilized to estimate the number of robots in the State of Michigan is that it directly assumes that on average the auto firms and other firms will install robots in Michigan in the same proportion as the relevant production worker employment in the state. The presumption is that production worker employment represents an adequate measure of the likelihood of robotizing Michigan's factories. That appears reasonable since robots will replace such workers, but it remains only a very rough approximation. In particular, decisions to robotize could be expected to reflect wage differentials and other determinants of production techniques. In addition, the rate of introduction of robots is expected to be more rapid where new plants are opened, thus reflecting choices of the location of new productive capacity. Influences such as these on the probability of robotization of Michigan factories cannot be accurately predicted at this time.

The derived forecast of the Michigan robot population can be found in table 2-12. Since the relative importance of the application areas *within* each industry remains the same as in the U.S. forecast, no discussion of those estimates is needed. The relative importance of the estimates *across* industries in Michigan, however, differs sharply from the U.S. totals. Specifically, three-quarters of the robots in Michigan are expected to be in the auto industry, while in the U.S. only about one-fourth of the robot population will be in autos.

In absolute terms, the number of robots in the auto industry in Michigan in 1990 is expected to range from a low of 5,327 units to a high of 8,879 units; the same figures for all other manufacturing in Michigan are 1,434 to 3,072 units. The combined total 1990 population of robots in Michigan is then 6,761 units to 11,951 units. Since about one-third of Michigan employment is in the auto industry, but three-fourths of Michigan's robots are expected to be applied there, it is obvious that the impacts will be much more dramatic in the auto industry.

The remainder of the monograph addresses the human resource impacts expected to result from this projected population of industrial robots in both the U.S. and in Michigan. The next chapter specifically addresses the question of job displacement, while the following chapter discusses those jobs that will likely be created by the spread of robotics technology. It will become clear as we proceed that the forecast of the robot population is the key link in our procedure. The robot forecast establishes the scale to

Table 2-12
Forecast of the Michigan Robot Population by Application, 1990

Application	Autos		All other manufacturing		Total	
	Range of estimate		Range of estimate		Range of estimate	
	Low	High	Low	High	Low	High
Welding	1,136	1,456	225	410	1,361	1,866
Assembly	1,492	3,125	205	614	1,697	3,739
Painting	639	888	131	225	770	1,113
Machine loading/unloading	1,776	2,841	717	1,393	2,493	4,234
Other	284	569	156	430	440	999
Total	5,327	8,879	1,434	3,072	6,761	11,951

which the specific employment impacts are adjusted. In our opinion, it is the consistency of these human resource impacts with the robot population forecast that is one of the major contributions of the study.

3
Displacement Effects of Robots

Before attempting to estimate the displacement effects of robots, it is important to insure that the meaning of the term "displacement" is clear. We use displacement to refer to the elimination of particular jobs, not to the layoff of individual workers. Certainly it is possible that the displacement of a particular job by a robot might lead to the layoff of the occupant of that job, but it is not necessary. Layoff refers to the involuntary separation of the worker from the firm, displacement refers to the elimination of the job itself without any assumption as to whether the worker in that job is separated from the firm, either voluntarily or involuntarily. Later in this chapter, after the discussion of the potential displacement effects of industrial robots in Michigan, the issue of unemployment resulting from this displacement will be discussed.

The basic methodology of this study proceeds from the forecast of the robot population presented in the last chapter. Once the number of robots by application area and industry can be specified (even within a broad range), it is only necessary to determine the average job displacement effect of each robot. Then these estimates of displacement by application and industry can be compared to the employment totals for the same application and industry to derive an estimate of the relative magnitude of job displacement associated with the projected robot population.

Average Rate of Job Displacement by Robots

Our interviews strongly supported the following conclusion about the average displacement effect of robots: one robot replaces one worker per shift. That conclusion should not be surprising. Robots are not any faster than human workers, and regardless of the protestations of some in the industry that robots should not be compared to humans, robots do in fact perform functions that were previously done by human workers. Engelberger admits that one focus of the development effort of the PUMA robot for small parts assembly was to make it human size to work alongside human workers. (Engelberger, p. 137) In several articles discussing cost justification of robots, John A. Behuniak, program manager of Automation Manufacturing Technology, General Electric Company, stresses that managers should guard against overly optimistic estimates of labor savings. (Behuniak, 1979 and 1981) He states, "Robots, unlike other forms of automation, usually only replace humans on a one-for-one basis." (Behuniak, 1979, p. 1)

There is a possibility that the average displacement effect may increase as technological improvements occur, as robot systems become more prevalent, and as the fund of human knowledge of robot applications increases. Tanner and Adolfson, in their study of the U.S. auto industry, conclude that one robot replaces 0.9 workers today in the auto industry, but that will improve to 1.2 workers by 1990. (Tanner and Adolfson, p. 103) Of course, these data relate to new installations only and not to the total stock of robots. So even if robot productivity improves, a substitution rate of one robot for one worker will not be far off.

It should be reiterated that stand-alone robot installations are expected to dominate over the next few years. By 1985 the UM/SME Delphi forecast anticipates only 20 percent of robot sales will be for inclusion in robot systems, and by 1990 that figure is expected to rise to 40 percent. (Smith and Wilson, p. 46) Currently it appears that the displacement effect of isolated robot systems or cells in production facilities is not much different from that of stand-alone robots themselves. However, as we slowly move toward the factory of the future and these cells are themselves linked together, some observers expect the displacement effect to rise dramatically.

On the other hand, it may be far too easy to overestimate the productivity (displacement) impact of technological change in general. Bela Gold, who has studied this question in many industries, concludes that even major technological changes have "fallen far short of their expected effects." (Gold, 1981b, p. 91) The source of the overestimate is the tendency to focus only on the change itself and thereby neglect the totality of the production process. (Gold, 1981b, p. 91) It is somewhat akin to recognizing the important difference between potential effects and actual effects, as discussed in chapter 1, and may in part account for the warning by Behuniak not to overestimate the labor savings attributable to robots.

There are several factors that will tend to mitigate the job displacement impact of industrial robots. First, as robots become more common in manufacturing processes, they will replace hard automation such as mechanical transfer devices, as well as human workers. This kind of substitution follows from the fact that industrial robots represent an intermediate technology between dedicated or hard automation and manual or human labor.

Second, as robots become more numerous, the need arises for redundancy in some robot installations. That is already occurring today in body assembly welding applications in the auto industry, where one or two robots at the end of the line are actually spares, available in the event of robot failure. Third, there will eventually be a need for replacement robots, although it is still too early to establish an average expected lifespan for a robot. Our interviews revealed estimates from 10 to 15 years; one person even maintained that with proper maintenance and replacement of parts the lifespan of a robot is indefinite, except where the work environment is unusually harsh, such as painting applications. The problem, of course, is that without specific information about discards, we may mistakenly count replacement robots as new robots and subsequently new displacement when in fact no new displacement has actually occurred. In any event, dramatic changes in the average displacement ratio are highly unlikely during the forecast period, and it is our judgment that the one-to-one relationship between robots installed and workers displaced represents the best approximation to the actual gross displacement impacts that will occur.

A closely related question is the number of shifts per day. This typically varies depending on the industry, stage of the business cycle, and sometimes seasonal factors. On the one hand, simple formulas suggested to evaluate robot acquisitions imply that single-shift operation of robots is not generally cost-effective, although some robots today are being so used. (Engelberger, pp. 104-106) On the other hand, most American manufacturing industry does not currently operate three shifts. The UM/SME Delphi forecast foresees little impact of robots on the number of daily shifts (Smith and Wilson, p. 82) but a careful reading of the rationale provided by the experts supporting their opinions reflects considerable disagreement on this issue. (Smith and Wilson, p. A-88) Some of the experts polled did point out that once robots are implemented in significant numbers, time must be allowed for robot maintenance. So industry movement to three-shift operation may not be likely even with widespread robot application.

Given the vagaries of product demand over the business cycle and direct maintenance requirements, it appears that an *average* number of shifts in excess of two is highly unlikely. But economic constraints appear to prevent widespread robot utilization in single-shift applications. Thus, in this study two-shift operations are assumed; so, on the average, two workers are displaced for each robot installed. As robot utilization becomes more common in production facilities, this simple ratio assumption may need to be re-examined, but with a 1990 focus two jobs displaced for each robot appears to be very reasonable.

U.S. Job Displacement

The robot population forecasts presented in the previous chapter can be translated directly into the number of jobs displaced on the basis of an average of two jobs per robot for each functional application. Table 3-1 reports these results for the U.S. as a whole. Clearly, the forecast of 50,000 to 100,000 robots operational in the U.S. by 1990 means that 100,000 to 200,000 jobs will have been displaced by robots during the decade of the 1980s.

Further, we expect job losses of 30,000 to 50,000 in the U.S. auto industry as a result of the application of robots, and 70,000 to 150,000 jobs lost in other manufacturing industries. These totals can be broken down to specific functional areas as well. For instance, our robot population forecast implies that 6,400 to 8,200 jobs will be eliminated by welding robots in the auto industry. Similarly, 11,000 to 20,000 welding jobs in other manufacturing industries will be lost by 1990.

Table 3-1
Estimate of Job Displacement in U.S. by Application, 1990

Application	Autos		All other manufacturing		Total	
	Range of estimate		Range of estimate		Range of estimate	
	Low	High	Low	High	Low	High
Welding	6,400	8,200	11,000	20,000	17,400	28,200
Assembly	8,400	17,600	10,000	30,000	18,400	47,600
Painting	3,600	5,000	6,400	11,000	10,000	16,000
Machine loading/unloading	10,000	16,000	35,000	68,000	45,000	84,000
Other	1,600	3,200	7,600	21,000	9,200	24,200
Total	30,000	50,000	70,000	150,000	100,000	200,000

A larger number of assembly jobs are expected to be eliminated by industrial robots during this decade. Table 3-1 reports that from 18,400 to 47,600 such jobs are at risk of robotization. The range of prediction for assembly displacement is wider than that for welding, owing to the uncertainties of robot assembly capability as discussed in chapter 2. If robots enjoy early success at more complicated assembly tasks, the job displacement will range to the higher end of our estimates. Painting robots will displace fewer jobs than either welding or assembly robots as shown in table 3-1. However, as we will see later, the relative impact on employment in this area looks to be very significant.

The greatest number of jobs will be eliminated by pick-and-place robots performing machine loading and unloading functions. These functions include casting, forging, press loading, machine tool loading and other similar operations. Table 3-1 suggests that roughly 40 percent of all robot job displacement will occur in this area. Machine loading and unloading is the best known general use of robots in U.S. industry today and will continue to be the most prevalent kind of industrial robot in the future. It is worthy of note that this area is significantly less concentrated in the auto industry than those discussed heretofore.

But these numbers have only limited meaning without reference to an employment base to put them in relative perspective. That is, it is interesting to know that there may be 3,200 to 4,100 welding robots in the auto industry by 1990, and that these robots can be expected to eliminate 6,400 to 8,200 jobs. But the impact of such a development depends to a considerable degree on the *relative* magnitude of this displacement. Does this represent 1 percent or 10 percent or 100 percent of the welding jobs? The answer to such a question is required before any conclusions about the seriousness of this situation can be reached.

For instance, if the welding jobs eliminated represent only a tiny fraction of welders employed in the auto industry, little disruption or distress would be expected. Normal employee turnover could be expected to effectuate the reduction in force required, and displacement of jobs need have no implications of layoff or unemployment. On the other hand, if a large proportion of jobs is represented, there is more cause for concern.

What is needed is an occupational data base organized by industry that makes possible the comparison of these potential job displacement figures with the existing employment levels in the same geographical area, function, and industry. Fortunately, such a detailed occupational data base does exist. It is called the Occupational Employment Statistics (OES) survey published by the U.S. Department of Labor, Bureau of Labor Statistics.

The OES survey was developed during the 1970s to fill the need for a relatively current, detailed data base for making occupational projections. The BLS had used industry-occupation matrices based on the 1960 and 1970 Decennial Censuses but found that a more frequent survey was desired. The 10-year intervals between observations just proved too infrequent to serve the purpose of projecting occupational needs. (U.S. Department of Labor, 1981b)

The OES survey is conducted jointly by the federal and state governments on a 3-year rotating schedule. All nonagricultural employers are divided into one of three groups according to industry. A sample of establishments from one of these groups is surveyed each year. Manufacturing employers were surveyed in 1971, 1974, 1977 and 1980, although coverage by state was rather spotty before 1977. (U.S. Department of Labor, 1980b, p. 91) Detailed occupational information is collected for a total of 1,678 occupational titles and 378 industries. The information is gathered from each employer using no more than 200 job titles that have been previously found appropriate to that industry. Employment is reported according to the highest skill level performed by an individual employee as of the 12th of April, May, or June, depending on which month shows the least seasonality in employment in the industry.

From this raw data, the BLS produces national industry-occupation matrices. In addition, most states participating in the joint effort also produce statewide matrices based on a common set of procedures developed by the BLS. To evaluate the *relative* magnitude of job displacement, we can compare the gross displacement estimates from table 3-1 to employment levels reported in 1980 in the OES survey of manufacturing employers.

This procedure should prove sufficient to put the magnitude of projected job displacement in perspective, but there are problems with misinterpreting the precision of such estimates. We discuss four of these problems. First, the utilization of employment data from any given year implies the assumption of constant output. Thus, using a 1980 employment data base to assess the significance of job displacement carries the assumption that output and employment in 1990 would be comparable to 1980 levels, except for the influence of robots. Of course, this is highly unrealistic; there are a multitude of forces that will cause output and employment levels in 1990 to differ from those in 1980.

The only alternative to this unrealistic fixed output assumption is to forecast the influence of all these other factors and then use the projected output and employment levels as a base for determining the relative impact of robots. Results of this type will be presented later in the chapter. Suffice it to say that there is some merit in using known facts as a basis to assess the significance of a change rather than a projection of unknown accuracy.

Second, in addition to the implicit assumption of fixed output, choice of a particular base year also carries with it the peculiar circumstances of that year. This can best be illustrated by recent employment trends in the auto industry. Auto sales and auto industry employment peaked in 1978. For the U.S. as a whole, employment in the auto industry declined precipitously from 1979 to 1980 and has shown no improvement since. In fact, 1982 employment levels reflect a further decline over the depressed 1980 and 1981 totals. Table 3-2 reports these totals with an index number showing the employment level relative to the 1978 peak for all employees and for production workers. It is clear from the table that employment in the auto industry has declined by roughly 30 percent from 1978 through May of 1982. Between 1979 and 1980 alone, the average total employment in the auto industry declined by over 20 percent, due to the impact of the recession and foreign imports.

Table 3-2
Employment in U.S. Motor Vehicle and Equipment Industry (SIC 371)
1978 to 1982

Year	Total employment (thousands)	Index (1978 = 1.00)	Production workers (thousands)	Index (1978 = 1.00)
1978	1,004.9	1.000	781.7	1.000
1979	990.4	.986	764.4	.978
1980	788.8	.785	575.4	.736
1981	783.9	.780	582.8	.746
May 1982	717.0	.714	533.2	.682

SOURCES: Data for 1978 through 1981 from *Supplement to Employment and Earnings: Revised Establishment Data,* Bureau of Labor Statistics, June 1982, p. 88. The 1982 data are from *Employment and Earnings,* Bureau of Labor Statistics, August 1982, Table B-2, p. 40.

But if one is trying to assess the relative displacement effect of robots, which employment level is appropriate? If the reduction in employment in 1980 is regarded as a permanent change, clearly 1980 is an adequate base year. If, however, the reduction is seen as a short term phenomenon, using 1980 employment numbers will significantly distort the results. Clearly, the reduction of 20 percent in auto employment in 1980 would cause an increase of 20 percent in the calculated displacement rate with the methods used here.

Actually, we would not regard either the peak 1978 employment levels or the depressed 1982 employment levels as a fair baseline. It is probably not reasonable to predict a return to 1978 employment totals in the auto industry, even if sales do recover to the 13 million level. The U.S. auto industry must raise the productivity of its labor force (or reduce the levels of compensation) if it is to meet the foreign competition. Thus it is appropriate to anticipate a declining labor input requirement for a given level of sales. The application of industrial robots is obviously one of the ways the industry is attempting to meet the challenge.

On the other hand, if profitability cannot be restored to the industry through significant sales gains in the short term, the capacity to finance the capital improvements (including robots) needed to meet the long term competitive goals will be seriously impaired. From the perspective of anticipated employment levels in the industry, these considerations lead us to believe that an employment base somewhere between the 1978 level and the 1982 level is most reasonable. The 1980 employment base utilized here thus appears overly pessimistic, but it represents the most recently available data from the OES base. Utilizing an employment base that underestimates the true level will serve to bias the job displacement rates upward. This issue will be discussed briefly again when the Michigan displacement figures are presented later in the chapter.

The third reason to be cautious about the precision of our estimates involves the definition of the auto industry itself. Up to now it has not been necessary to specify precisely what is meant by the term auto industry. Implicitly we have used the term to refer to the major auto producers, General Motors, Ford, Chrysler and American Motors. This is primarily a matter of convenience, but also reflects the judgment that robots will continue to be large firm technology through at least 1990. At the other definitional extreme would be what the Michigan Employment Security Commission has called "Motor-Vehicle-Related" employment. (Michigan Employment Security Commission, 1981b, p. 2) This includes not only the assembly of motor vehicles, but component parts suppliers, raw material providers, and tool and die shops as well.

Unfortunately, neither of these options are workable when matching up against an occupational data base. The OES uses three-digit SIC codes as the minimum level of aggregation available to the public. This poses a potential problem of comparability with the robot forecasts presented earlier. The only workable definition of the auto industry at the three-digit SIC level is SIC 371, Motor Vehicles and Equipment. This will include auto parts and accessory manufacturers, but excludes stamping plants, engine plants, and other major component manufacturing from the industry-occupation data base. The result is that there can be a discrepancy between the areas of application of the robots and the occupational employment figures against which the displacement should be measured. Given the lack of more specific information, these inaccuracies must be tolerated.

The final reason to question the precision of our estimates is the problem of determining what, if any, displacement impacts have already been registered in the U.S. economy. Ideally, an employment observation dating before any substantial robot deployment would eliminate the possible question of intermediate impacts. That is, if the employment profile available predates the application of robots, there is no necessity to try to determine what effects have occurred to date and are therefore already imbedded in the measured employment base. Unfortunately, there is insufficient information to determine the industry of use or application areas for the less than 2,000 robots installed in the U.S. by the time of the survey. Thus, we must disregard these intermediate impacts, which are quite small in any event, when using a 1980 employment base.

In summary, there are many reasons to be wary of the precision of our displacement rates. There are serious questions about the assumption of constant output, the appropriateness of the base year, the definition of the auto industry, and the neglect of intermediate impacts. Thus the estimates of the relative magnitude of job displacement must be regarded as representing a *general range* rather than a precise point. We do believe, however, that we *can* identify the general order of magnitude of robot impacts, even if specific estimates are inaccurate. Further, we submit that it is the order of magnitude that should shape any policy response to the challenge of robotics at this early date.

Table 3-3 presents the estimated displacement impact of industrial robots in the U.S. Since employment levels are from 1980 and the robot population forecast is for 1990, the job displacement is a cumulative total over the 1980 to 1990 period. For each of the application areas listed, a specific occupational correlate was selected from those available in the OES. For the welding robot application area, the occupational title "welders and flamecutters" was selected as representing the job content against which the projected number of welding robots would be applied. For assembly robots, the "assembler" titles were judged to be the employment base. For painting robots, the "production painter" occupation in the OES was chosen. For the machine loading

Table 3-3
Displacement Impact of Robots in the United States by Application, Cumulative 1980 to 1990

Application	Autos		All other manufacturing		Total	
	1980 employment level	Displacement range (percent)	1980 employment level	Displacement range (percent)	1980 employment level	Displacement range (percent)
Welding	41,159	15 - 20	359,470	3 - 6	400,629	4 - 7
Assembly	175,922	5 - 10	1,485,228	1 - 2	1,661,150	1 - 3
Painting	13,556	27 - 37	92,622	7 - 12	106,178	9 - 15
Machine loading/ unloading	80,725	12 - 20	988,815	3 - 7	1,069,540	4 - 8
All operatives and laborers	467,846	6 - 11	9,954,048	1 - 2	10,421,894	1 - 2
All employment	773,797	4 - 6	19,587,771	0 - 1	20,361,568	0 - 1

SOURCE: Employment data based upon unpublished OES data provided by Office of Economic Growth and Employment Projections, Bureau of Labor Statistics, U.S. Department of Labor, Washington, DC.

and unloading applications, the semi-skilled metalworking operative group (with the exclusion of welders and flamecutters) was selected. This includes drill press operatives, grinding and abrading machine operatives, lathe and milling machine operatives, punch and stamping press operatives, and other precision machine operatives.

Clearly, the choice of the specific occupational content against which to apply the robot displacement figures is somewhat arbitrary. For instance, we chose to apply the painting robots against employment in the production painter occupational category. But it is quite likely that some of the jobs actually displaced will come from other occupations, say, materials handlers or general laborers. This is particularly likely if a robot system is installed rather than just a stand-alone retrofit robot. Again, the conclusion is that these numbers should be taken as indicative of the general range of impact on occupational employment.

Table 3-3 indicates that the robot population forecasts presented in chapter 2 will have widely varying impacts on different occupations. The most dramatic displacement rate is for production painters in the auto industry. Our results show that from 27 to 37 percent of these jobs may be eliminated by 1990. In addition, table 3-3 shows that 7 to 12 percent of production painter jobs in other manufacturing industries may be displaced. Overall, 9 to 15 percent of these jobs are threatened by robots in this decade. These results should not be surprising. Painting is a prime robot application, in that it is based on existing technology, and painting itself is a particularly dirty and potentially hazardous job.

Significant job displacement is also anticipated for welding occupations, another prime robot application. Table 3-3 indicates that 15 to 20 percent of welder jobs in the auto industry will be displaced by welding robots. A lesser impact on other manufacturing welding jobs is also shown, with 3 to

6 percent displacement forecast. For all manufacturing, 4 to 7 percent of welder jobs are expected to be eliminated by 1990.

Very similar results apply to the machine loading and unloading robot application area. Overall, 4 to 8 percent of this large employment group can expect to be displaced by 1990, with autos showing three to four times the impact of other manufacturing. Since this job content is already automated to some degree, robot application here calls for integrating the robots into the existing production facilities. This is notoriously difficult to accomplish smoothly.

This is even more true for the assembly robots, the task which shows the least impact in table 3-3. While 5 to 10 percent of auto assembly jobs are expected to be robotized, only 1 to 2 percent of such jobs in other manufacturing are at risk. The aggregate manufacturing displacement rate for assembly will only be 1 to 3 percent. As mentioned earlier, this impact depends on continued refinements in robot capability, adaptability, and reliability. The projections here are clearly more speculative than those for welding or painting.

Some may question why some of the specific occupational displacement rates are not higher, particularly in the auto industry. For example, in chapter 2 it was pointed out that the installation of welding robots in the auto industry will slow down after 1988. Presumably that represents maximum application of robotics technology to welding functions. Why, then, isn't the displacement of welders approaching 100 percent?

There are a number of reasons for this apparent discrepancy. First, it is clear that the OES occupation of welders and flamecutters in the auto industry includes people doing work other than welding auto bodies together on the assembly line, the primary robot application. Second, the auto industry as defined in the OES data base includes many small firms producing auto parts and accessories. As discussed earlier, we do not expect small firms to adopt robotics technology in substantial numbers before 1990. So there is a considerable population of welders in smaller firms who may not even see a welding robot by 1990.

Third, even though we have tried to be as careful as we can in formulating our estimates and applying them to the available occupational data base, there are bound to be many inaccuracies associated with such a procedure. For instance, actual job classification schemes utilized in most large firms tend to be much more detailed than the occupational data available in the OES data base. Thus when we think of occupational displacement, we may be utilizing a broader definition of an occupation than some other observers.

In general, without a detailed case study at the plant level of labor input vectors before and after a specific technological change, it is difficult to say how much our simplifying assumptions may have influenced particular occupational displacement rates. This is yet another reason to regard our relative displacement rates as estimates of the general order of magnitude of job elimination rather than precise point estimates. It is also one of the reasons that we will next assess the impacts of robots against more general measures of employment, where occupational classification is not a factor.

The two bottom rows of table 3-3 express the relative displacement impact in this more aggregated manner. The job displacement expected is calculated as a proportion of all employment and of all operatives and laborers, in other words, the semi-skilled and unskilled manufacturing labor force. As shown in the table, aggregate job displacement of 4 to 6 percent in the auto industry is anticipated. However, this represents 6 to 11 percent of operatives and laborers, the "blue collar" workforce. Thus, during the decade of the '80s production worker employment in the auto industry is projected to be up to 11 percent lower than it otherwise would be due to the introduction of robots.

Table 3-3 also reveals that anticipated impacts are much lower in other manufacturing industries. From 1 percent to 2 percent of all semi-skilled jobs will be eliminated by 1990 in these areas. Of course, individual industry impacts would tend to be higher than the average in those industries where robots are well suited to production techniques. On the other hand, we are not aware of any other industry that will show a gross impact equal to the auto industry. Thus it is not unreasonable to suppose that our displacement figures for the auto industry represent an upper bound for other manufacturing industries during the forecast period.

In summary, table 3-3 indicates that robots will *not* have a significant direct impact on overall employment levels in the U.S. between now and 1990. Robots in total will eliminate less than 1 percent of all jobs in this period. In the auto industry, however, overall displacement impact does appear significant. In particular applications like welding and painting, the job displacement impact is quite dramatic. The importance of these job displacement results is that they indicate a wide range in the impact of robotics. While there is no cause for concern in an aggregate sense, there will be pockets of significant job displacement. Later in the chapter we will attempt to describe the possible unemployment impact of this job displacement.

Before going on to discuss the potential displacement induced by robots in Michigan, it may be useful to compare our U.S. results with others, especially the Carnegie-Mellon study and the UM/SME Delphi forecast. Table 3-4 presents these comparative results organized according to the applica-

Table 3-4
Comparison of Displacement Rates, Various Studies

Occupation	W. E. Upjohn Institute estimates		Carnegie Mellon survey	UM/SME Delphi Forecast[b]	
	Auto	Total	(Level 1)[a]	Potential	Actual
Welding	15 - 20	4 - 7	27	20	10
Assembly	5 - 10	1 - 3	10	10	5
Painting	27 - 37	9 - 15	44	20	15
Machine loading/unloading	12 - 20	4 - 8	20	10	6

a. The displacement rates shown are those from the weighted average response for Level 1 robots. See Robert Ayres and Steven Miller, *The Impacts of Robotics on the Workforce and Workplace,* Carnegie-Mellon University, Department of Engineering and Public Policy, June 1981, pp. 97-99.

b. From Donald N. Smith and Richard C. Wilson, *Industrial Robots: A Delphi Forecast of Markets and Technology,* Society of Manufacturing Engineers, Dearborn, Michigan, 1982, p. 70.

tion areas used for this study. Maximum comparability was sought here, but comparisons across studies must always be interpreted cautiously because of differing scope, puposes, methods, definitions, time periods, etc. Even the occupations themselves are not identical across the studies, although they are similar. In general, our estimates of displacement are the lowest ones shown. This is especially marked when comparing our total displacement rates with those from the Carnegie-Mellon study. But the results from the UM/SME Delphi forecast may help make it clear why this is so.

Recall from chapter 1 that the Carnegie-Mellon study asked what proportion of the work done by given occupations *could* be performed by Level 1 or Level 2 robots. Thus the question corresponds most closely with the *potential* displacement figure from the UM/SME Delphi survey. In fact, the Carnegie-Mellon displacement rates are even higher, though the rates shown are the average response for Level 1 robots only. It is also interesting to note that the Carnegie-Mellon displacement rates are rather close to the top end of our range for the automobile industry.

This result may be due in part to the Carnegie-Mellon weighting procedure which gives greater importance to large employers. The large firms in their survey are more likely to be auto or auto related firms, since there is a disproportionate concentration of both robot users and large establishments within the auto industry. (Ayres and Miller, 1981a, p. 100) This raises the possibility that the Carnegie-Mellon displacement estimates could be more descriptive of the auto industry than an all-industry average, at least through 1990. Our total manufacturing gross displacement rates agree more closely with the UM/SME Delphi survey *actual* expected displacement rates than with the UM/SME Delphi potential rates. This is to be expected since our projections are also targeted on the actual rather than theoretical or potential impacts. In fact, our estimates tend to be slightly lower than the UM/SME Delphi estimates of actual expected displacement.

In summary, the Carnegie-Mellon displacement results are quite similar to our projections for the auto industry. Given that the auto industry is the leader in the application of robots, this may corroborate their theoretically possible levels of displacement for Level 1 robots. We do not feel, however, that their displacement results can be generalized across all manufacturing industries. Our gross displacement rate estimates coincide much more closely with the results of the UM/SME Delphi forecast. We are not dismayed by the fact that our estimates tend to be somewhat more conservative than even the UM/SME Delphi actual projections. We discussed the reasons for our reservations extensively in chapter 2. We repeat our belief that most of the observers of robotics are erring on the side of technical progress without full consideration of the human, organizational, and financial limits to changes in process technology.

Michigan Job Displacement

For the State of Michigan, OES surveys of manufacturing employers were conducted in 1977 and 1980. However, the 1980 survey data are not yet available for public release, so the 1977 survey is actually the only such detailed data base currently available for the State of Michigan. (Michigan Employment Security Commission, 1981a) Actually, we will be using a BLS standardized 1978 update of these raw 1977 Michigan numbers to maximize consistency of the information. These unpublished data were provided by the MESC at the three-digit SIC level, in contrast to the published 1977 data which are at two-digit industry level only.

Table 3-5 shows the estimates of job displacement resulting from the application of robots in the State of

Table 3-5
Estimate of Job Displacement in Michigan by Application, 1990

Application	Autos		All other manufacturing		Total	
	Range of estimate		Range o estimate		Range of estimate	
	Low	High	Low	High	Low	High
Welding	2,272	2,912	450	820	2,722	3,732
Assembly	2,984	6,250	410	1,228	3,394	7,478
Painting	1,278	1,776	262	450	1,540	2,226
Machine loading/unloading	3,552	5,682	1,434	2,786	4,986	8,468
Other	568	1,138	312	860	880	1,998
Total	10,654	17,758	2,868	6,144	13,522	23,902

Michigan. It is based on the Michigan robot forecast presented in chapter 2 and the assumption of two jobs displaced per robot applied. Overall, we project that between 13,522 and 23,902 jobs will be displaced in Michigan by 1990. Because of the structure of employment by industry in the State of Michigan, the impact of robots in the auto industry looms much larger than for the U.S. as a whole.

Table 3-5 shows that 75 percent of the job loss in Michigan is expected to be in the auto industry (SIC 371) with gross displacement of 10,654 to 17,758 auto jobs. Nearly 2,000 painting jobs, 3,000 welding jobs, 6,000 machine tending jobs, and over 6,000 assembly jobs in the auto industry could be lost to robotization by the end of the decade in Michigan. The results in the table also show that gross job displacement in Michigan will be very minor outside the auto industry.

Table 3-6 presents these same job displacement results expressed in relative terms. Each gross job displacement figure in table 3-5 is divided by the corresponding occupational employment from the OES for Michigan in 1978. Thus the displacement rates presented in table 3-6 represent the cumulative total gross displacement proportion for that occupational group over the period from 1978 to 1990.

In addition to these specific occupational rates, the bottom lines of the table show the overall displacement rates calculated against all employment and against employment in the operative and laborer sectors, generally encompassing the semi-skilled and unskilled workers. From the table it is apparent that the projected Michigan robot population in 1990 will displace somewhere between 1 and 2 percent of all 1978 manufacturing jobs in the state. When assessed against only the semi-skilled and unskilled employment base, the proportion of job displacement exactly doubles to between 2 and 4 percent. In both instances, the Michigan displacement impact is roughly twice that of the U.S.

Table 3-6
Displacement Impact of Robots in Michigan by Application, Cumulative 1978 to 1990

Application	Autos		All other manufacturing		Total	
	1978 employment level	Displacement range (percent)	1978 employment level	Displacement range (percent)	1978 employment level	Displacement range (percent)
Welding	14,910	15 - 20	22,694	2 - 4	37,604	7 - 10
Assembly	65,764	5 - 10	50,678	1 - 2	116,442	3 - 6
Painting	4,378	29 - 40	4,387	6 - 10	8,765	17 - 25
Machine loading/ unloading	42,149	8 - 14	86,906	2 - 3	129,055	4 - 7
All operatives and laborers	206,927	5 - 9	397,598	1 - 2	604,525	2 - 4
All employment	409,506	3 - 4	769,841	0 - 1	1,179,347	1 - 2

SOURCE: Employment data based upon unpublished OES data provided by Office of Economic Growth and Employment Projections, Bureau of Labor Statistics, U.S. Department of Labor, Washington, DC.

Outside the auto industry, robot job displacement in Michigan during the decade of the '80s will be minimal, with one exception. It appears likely that production painters in all other manufacturing will experience significant displacement. Table 3-6 indicates that the application of painting robots can be expected to eliminate 6 to 10 percent of these production painter jobs by 1990.

The significant job displacement in Michigan will be concentrated in the auto industry. This results from Michigan's dependence on the auto industry and the circumstance that the auto industry will continue to lead other industries in robotization. According to table 3-6, from 3 to 4 percent of Michigan 1978 auto industry employment can be expected to be eliminated by industrial robots by 1990. When the displacement is expressed as a percentage of only the semi-skilled and unskilled labor component, the rate rises to 5 to 9 percent.

When attention is confined to the prime robot applications, once again it is seen that 15 to 20 percent of the welding jobs in the Michigan auto industry are expected to disappear by 1990. For production painters the proportion is even higher, from 29 to 40 percent eliminated. These must be deemed significant job displacement rates by anyone's standards. They will in all probability create some labor market dislocation in Michigan in the absence of some intervention, either private or public. Furthermore, these impacts can be predicted with lower ranges of uncertainty because the technology is already known. In these applications it is primarily a question of the rate of diffusion of currently existing techniques.

In summary, the Michigan displacement estimates are similar to those of the U.S. Robots will not have an enormous impact on overall employment levels in the State of Michigan between now and 1990. Robots are projected to have no significant negative impact in nonauto manufacturing with the possible exception of production painters. In the auto industry, however, overall displacement does appear significant; it is quite dramatic in particular applications like welding and painting.

Before moving on to the question of possible labor market implications, there are two other issues which must be dealt with. The first is the question of the effect of the 1978 employment base on the displacement figures reported in table 3-6. Outside the auto industry, there appears to be no problem because of the relative stability of manufacturing employment. But, as mentioned earlier, 1978 represented a cyclical peak for the auto industry. Indeed, if one compares table 3-3, the U.S. estimates of displacement, with table 3-6, the Michigan estimates of displacement, these points are illustrated quite well. The displacement rates are generally similar for all other manufacturing in both cases. The aggregated displacement rate in the auto industry for all operatives and laborers, however, is 20 to 25 percent less in Michigan than the U.S., reflecting the effect of the 1978 cyclical employment peak. The all employment figures for the auto industry cannot be compared because of the relative concentration of automotive adminstrative headquarters and research facilities in Michigan. Likewise, the total estimates cannot be compared because of the rather significant difference in industrial structure between Michigan and the nation as a whole. In short, at the aggregate level the utilization of the 1978 employment base for the Michigan auto industry estimates appears to confirm our expectations of the effect of using a different employment base.

However, it is puzzling that the specific occupational displacement rates in the auto industry in Michigan are *not* generally lower than the corresponding U.S. estimates. This puzzle may be explained by recalling the method used to derive the Michigan robot population. Production worker employment in the auto industry in Michigan relative to the nation as a whole was assumed to adequately reflect robotization potential. Thus, a gross production employment figure was used to assign robots to the State of Michigan. Presumably the unique structure of auto employment in Michigan was not captured by this procedure.

For these reasons we think the 1980 estimated displacement rates for the U.S. auto industry may be more meaningful even for Michigan, although it bears repeating that the aggregate employment levels in the auto industry in 1980 appear overly pessimistic. Thus the U.S. job displacement results may provide an upper bound for our estimates.

The other major question is the location of the jobs displaced within the State of Michigan. While there are no data available that would make is possible to accurately represent the occupational profile of sub-state regions, it is reasonable to assume that the job displacement will occur throughout the industry. The best assumption, absent a major effort to delineate exactly what job content is present in each auto-related establishment in Michigan, is simply to assume that the job displacement will occur where the current production worker jobs are located.

Table 3-7 is adapted from MESC data collected for a study of Michigan's auto dependency. It shows that, in March of 1979, 60 percent of the "motor-vehicle-related" employment in the state was located in the Detroit SMSA. An additional 17.6 percent was located in the outlying Flint and Ann Arbor-Ypsilanti SMSAs. If the 9.4 percent accounted for by Saginaw and Lansing-East Lansing is added to these numbers, 87.0 percent of the auto employment is in the southeast Michigan quadrant. Accordingly, we would assume that 87 percent of the job displacement resulting from the application of robots will occur in Southeast Michigan as well.[1]

1. This situation has not changed since 1979. In 1981, 88.1 percent of employment in SIC code 37, transportation equipment, was in Southeast Michigan.

Table 3-7
Motor-Vehicle-Related Employment in Michigan SMSAs, March 1979

SMSAs	Motor-vehicle-related employment	
	Number	Percent
Detroit SMSA	393,100	60.0
Ann Arbor-Ypsilanti SMSA	36,300	5.5
Flint SMSA	79,100	12.1
Lansing-East Lansing SMSA	32,300	4.9
Saginaw SMSA	29,200	4.5
All other areas	85,200	13.0
Michigan total	655,200	100.0

SOURCE: Adapted from *Motor Vehicle and Related Industries in Michigan,* Michigan Employment Security Commission, Bureau of Research and Statistics, Summer 1981, Table VI, p. 14.

Anticipated Impact of Job Displacement

Having completed the discussion of which jobs are likely to be displaced by robots by 1990, it is time to turn to the more critical issue of the possible unemployment impact of the elimination of up to 200,000 U.S. jobs in this decade. That is, how much labor market distress is likely to result from the job elimination which has been described here?

The first point that should be made is that we do not believe that the impact of robotics can truly be separated from other forces influencing employment levels between the present and 1990. However, for the purpose of assessing the possible unemployment impacts of robotics, we will examine this one development in isolation, as if it were the only change. Once again our purpose is to affix the order of magnitude of job elimination due to robotics relative to other (more ordinary) labor market developments, not to reach precise estimates of the impact of robotics on the unemployment rate.[2]

While the gross displacement results presented thus far appear to give relatively little cause for concern except in the auto industry, a fuller appreciation of the level of projected job displacement can be obtained from table 3-8. This table compares the simple average annual job displacement rate over the 1980 to 1990 period from our projections with average annual replacement needs and total openings as estimated by the Bureau of Labor Statistics, Office of Economic Growth and Employment Projections.

The BLS projects employment levels and demand for labor as a part of their labor market information system to assist program planners and individual decisionmakers in career choices. As reported in table 3-8, they also derive estimates of average annual replacement needs and total average annual openings. Average annual replacement needs are those job openings due to deaths, disabilities, and retirements only, while total average annual openings also include the BLS projected change in demand for that occupation. Neither data series, however, includes occupational transfers, i.e., the extent to which people voluntarily change occupations; so the relative rates in the table understate true annual labor market needs or vacancies.

The last column in table 3-8 shows that for welders, the BLS projects that job openings will average 5.1 percent annually over the period 1978 to 1990. Further, replacement needs alone, without any expansion in welding employment, would require filling 2.3 percent of all welder jobs every year. As the first three columns show, this is far above the

2. For an interesting account of a much more elaborate input-output version of this type of exercise in assessing the impact of technological change on the Austrian economy, see the article by Leontief listed in the bibliography.

Table 3-8
Displacement Impacts of Robots
Compared to BLS Estimates of Job Openings

	Simple average annual displacement impact of robots 1980 - 1990*			BLS average annual replacement needs 1978 - 1990	BLS total average annual openings 1978 - 1990
Application	Autos	All other manufacturing	Total	All industries	All industries
Welding	2.0	.6	.7	2.3	5.1
Assembly	1.0	.2	.3	3.0	6.5
Painting	3.7	1.2	1.5	2.4	3.9
Maching loading/ unloading	2.0	.7	.8	2.5	3.0
All operatives and laborers	1.1	.2	.2	2.9	4.0
All employment	.7	.1	.1	3.8	5.5

SOURCE: Replacement needs and total average annual openings from *The National Industry-Occupation Employment Matrix, 1970-1978, and Projected 1990,* U.S. Department of Labor, Bureau of Labor Statistics, Bulletin 2086, Vol. 2, April 1981, pp. 495-502.

*Assuming maximum growth in robot population.

annual displacement rate for all manufacturing industries estimated earlier. It is even slightly above the annual job displacement rate projected in this study for welders in the auto industry.

These BLS projections were made prior to any significant sales of robots, so it is unlikely that the BLS made any specific allowance for robots in their occupational forecasts. It is also important to note that the actual job content of each occupational category in table 3-8 is not identical to those used in this study. The BLS 1978-1990 forecast used the 1970 Census of Population as a base; hence it employed the Census Bureau occupational classification structure. This scheme is considerably less detailed than the OES base underlying our displacement projections.[3] But with the caveat that it is the most comparable data we can get, analysis of job openings and job displacement can be very illuminating. Any distortions should be small in most cases because we are looking at relative rates rather than absolute levels.

In no case do our job displacement rates exceed the BLS average annual openings figure. In fact, our manufacturing displacement numbers do not even come close to the replacement needs for all industries except in the case of painters. Whereas the BLS forecasts total job openings of 3.9 percent per year and replacement needs of 2.4 percent per year, our displacement rate ranges as high as 1.5 percent annually for painters in all manufacturing. If both forecasts were accurate, the introduction of industrial robots could be expected to eliminate roughly 62 percent of the painter jobs that would be opened through death, disability, or retirement. The point is, there still would be job openings for painters available each year, as well as additional vacancies due to occupational transfers.

3. For a comparison of the two occupational classification schemes, see U.S. Department of Labor, 1981b.

Using the same reasoning, our estimated robot job displacement impacts will eliminate about one-third of the welding and machine loading and unloading job openings due to replacement needs. Only about one-tenth of replacement assembly jobs would be eliminated. Even less than one in ten replacement positions for all operatives and laborers will be affected. Thus, if our most optimistic robot forecast proves accurate, the net result is that about 5 percent of projected annual job openings for operatives and laborers through 1990 will not materialize. About 7 percent of labor replacement needs for this group will be nullified. This is not a trivial result, but it is also not the end of the world for the blue collar worker.

There is another way to use the BLS occupational forecasts to illuminate the magnitude of job displacement by robots. One can compare the BLS projected employment growth by occupation to 1990 with the gross displacement projections by occupation reported earlier in this chapter. In this case, the occupational classifications are identical since the BLS results are also built upon the OES data base. This approach has the further significant advantage of eliminating the unrealistic assumption of constant output which was implicit in our displacement rates calculated on today's known employment base.

In accord with the economic assumptions behind the low-growth and high-growth variants of the BLS occupational employment projections,[4] we have deducted our gross job displacement figures from the corresponding employment projection. Thus we subtracted our low robot growth displacement figures from BLS low employment growth projections and similarly our high-growth forecast displacement from their high-growth projections. This approach is

4. For more information about the BLS 1990 projections, the interested reader should consult *Monthly Labor Review,* October 1981 (U.S. Department of Labor, Bureau of Labor Statistics).

Table 3-9
Projected U.S. Employment Changes 1980-1990 Utilizing Bureau of Labor Statistics Estimates Adjusted for Displacement Caused by Robots

Application	Autos			All other manufacturing			Total		
	1980 employment level	Percent change 1980-1990 Low growth	Percent change 1980-1990 High growth	1980 employment level	Percent change 1980-1990 Low growth	Percent change 1980-1990 High growth	1980 employment level	Percent change 1980-1990 Low growth	Percent change 1980-1990 High growth
Welding	41,159	3	15	359,470	17	29	400,629	15	28
Assembly	175,922	14	25	1,485,228	18	28	1,661,150	18	28
Painting	13,556	-8	-2	92,622	15	22	106,178	12	19
Machine loading/unloading	80,725	6	15	988,815	14	25	1,069,540	14	24
All operatives and laborers	467,846	12	24	9,954,048	11	19	10,421,894	11	20
All employment	773,797	15	29	19,587,771	14	23	20,361,568	14	23

SOURCE: Employment data based on unpublished OES information provided by Office of Economic Growth and Employment Projections, Bureau of Labor Statistics, U.S. Department of Labor, Washington, DC.

reasonable because the economic assumptions of the BLS correspond closely with our economic assumptions stated in chapter 2. Table 3-9 reports the result.

In each case, the 1980 employment in the occupational group is reported, together with the net employment change from 1980 to 1990 under both the low-growth and high-growth variants. Last, the percentage change in employment from 1980 to 1990 is reported. Frankly, this procedure attributes too much accuracy to both forecasts, but it is an interesting exercise that helps to put the job displacement effects of robots in perspective.

What table 3-9 tells us is that our job displacement totals are not sufficient to offset BLS predicted expansion of employment for any impacted occupation, except in the case of production painters in the auto industry. This is the only declining employment cell in the table. Even welders in the auto industry are forecast to expand in numbers during the 1980 to 1990 decade. In fairness to the BLS, it should be made clear that these forecasts were completed before the depth and breadth of the current auto slump were apparent. We would not regard the results of this exercise as serious forecasts of 1980 to 1990 employment change in light of more recent developments. But table 3-9 does put the job loss projected to result from the application of industrial robots in perspective. It will have little influence on employment trends to 1990 except in highly specialized situations.

Another analysis of the labor redundancy issue has been done in an unpublished General Motors Institute thesis by Jeffrey Krause. Using the announced General Motors target of 14,000 robots installed by 1990 and a displacement of two jobs per robot, Krause finds 28,000 GM workers potentially displaced by 1990. On the other hand, he cites a projected natural attrition rate for GM hourly employees through 1990 of 4.1 percent annually. (Krause, p. 104) Applying the stated attrition rate to the 1981 hourly labor force, Krause finds that a predicted turnover of 97,000 GM employees by 1990 is implied. His conclusions were: "First, by itself, robotics will contribute a relatively small amount to the overall reduction in the workforce; second, the rate of employee displacement due to new robotic applications will be gradual, relative to the rate of natural attrition." (Krause, p. 104)

Krause goes on to discuss other possible influences on the employment levels at General Motors, concluding with the following statement: "The displacement of 28,000 workers in General Motors should be compared to the approximately 140,000 workers presently laid off, due to lagging sales, poor economy, and intense foreign competition." (Krause, p. 105)

It would seem that even for the auto industry, the overall displacement induced by robot applications by 1990 will not be a major problem. Only for specific occupations within the auto industry, those particularly amenable to robotization with current technology, will job displacement be likely to cause significant labor market dislocations.

This result is also reflected in the UM/SME Delphi survey. When the Delphi survey sample was asked about the expected disposition of the workers displaced by robots during the 1980s, they responded with the following results. (Smith and Wilson, p. 75) Twenty-five percent of the workers were expected to be transferred to other jobs without additional training. Fifty percent were expected to be retrained for new positions in the same plant. Thirteen percent were expected to be retrained for new positions at another plant of the same employer. A total of 6 percent of displaced workers were expected to be terminated. An additional 6 percent were expected to be retired early. Thus, a maximum of 12 percent of all displaced workers were expected to be separated from their current employers.

The comparisons so far presented have been made on the basis of simple annual averages and decade long impacts. One might plausibly argue that this approach overstates the displacement impact of robots early in the decade but understates it later in the decade when sales of robots become much more significant. That is true in the aggregate, but not necessarily for individual occupations. The reason is that some robot applications, such as welding and painting, are expected to be much more important early in the decade. Others, such as assembly, are expected to rise in importance later in the decade. Furthermore, the timing of robot purchases on an annual basis by application area is even more uncertain than the total cumulative plans themselves, and there is no such data base available in any event. For these reasons we conclude that the simple annual averages by occupation such as those in table 3-8 are the most meaningful available.

However, we can determine the aggregate displacement impact of robots in 1990 alone. We estimated maximum sales of robots of 28,350 units in that year. The aggregate displacement effect of this maximum level of sales in 1990 is .3 percent of all 1980 manufacturing employment or .5 percent of all 1980 operatives and laborers. These single-year gross displacement impacts are also much less than replacement needs. If the BLS replacement numbers reported in table 3-8 are accurate, the 1990 single-year job displacement effect would eliminate roughly one blue collar job opening out of eight; one out of six for replacement openings. While this single-year impact appears more significant, it is still not cataclysmic by any means.

From at least four different perspectives then, the magnitude of *worker* displacement appears even less significant than the magnitude of *job* displacement presented earlier in the chapter. Even in the areas of most dramatic impact, the best evidence is that robots alone will not generally be enough to offset projected growth in employment. Job displacement levels to be produced by robots in this decade are small even when compared to replacement requirements of the labor force. Thus, on the basis of the evidence presented here, it appears that we will continue to need welders, machine operators, and assembly workers for the immediate future, even in the auto industry.

From the broad perspective, it is apparent that the rapid spread of robotics technology through American industry will not throw any significant number of American workers out of their jobs in this decade. It may still be true in 1990, as is claimed by industry sources today, that "no worker in American has lost his or her job because of a robot." The conclusion of this examination of job displacement by robots and the possible unemployment implications is that robots will have very little influence on aggregate levels of employment and unemployment in the decade of the 1980s. However, this conclusion must be tempered somewhat by a number of observations.

First, it is important to point out that robots could still add to the unemployment problem, even if no one actually loses their job due to a robot. If jobs that would have been created in the absence of robots are not created, there is a loss in the demand for labor, a loss in the total number of jobs. In these circumstances, it seems likely that the burden of unemployment generated by robots, if any, will fall on labor market entrants. Those who have not secured an entry to the factory before the robots arrive may be excluded. Thus robotics will play a role in the continuing loss of job opportunities for the relatively unskilled worker. This is part of a process of substitution of machines for humans in production that began over 200 years ago. The application of robots is yet another step in this evolutionary process.

It should also be noted that job elimination can have positive implications too. Robotics technology is generally being applied first to dirty, dangerous work situations. One of the guidelines for robot installations is, "look for the concentration of workers' compensation claims arising for prime robot applications." So there can be gains in social welfare associated with replacing humans in hazardous occupations with machines. This may even be true of the majority of robot installations in the U.S. to date. However, as robotics technology diffuses more widely, such as to assembly applications, the elimination of dirty and dangerous tasks will probably become a less important criteria for specific applications.

Last, it is clear from the evidence presented here that the job elimination impacts of robots will not be evenly spread occupationally, industrially, or geographically. We have shown that production painters will be most heavily impacted in the next few years. We have demonstrated that the auto industry will experience much more job elimination than the average manufacturing industry. We have also examined the potential impact on a single state and found that the job displacement impact of robotics will be concentrated in the southeast quadrant of Michigan. There will be other such pockets of localized impact where current employment is concentrated in manufacturing areas particularly susceptible to robots technology. So even if the overall unemployment implications of robotics are negligible, there will still be specific sites or specific occupations where the impact may be significant in this decade. Thus, there could be a displaced worker problem in such areas even if there is no general problem.

We will return to these displacement issues again in the conclusions chapter when policy implications of the study are discussed. Let us turn now to the other side of robotics technology: the jobs that will be created as robots spread through American industry, and the requirements for worker training or retraining that those jobs will impose.

Only after this side of the picture is fully discussed will the final conclusions about job displacement and potential unemployment emerge.

4
Job Creation

Introduction

Currently there are only scattered general statements about the job creation potential of robotics. A study by the Joint Economic Committee of the U.S. Congress concluded that "even if the most optimistic forecasts of sales growth materialize, total employment in robotic manufacture would not exceed 50,000 at any time in the next decade." (Vedder, p. 24) The UM/SME Delphi forecast estimated that 70,000 to 100,000 robotics-related jobs would be created by 1990. (Smith and Wilson, p. 67) The Carnegie-Mellon study indicated that the job creation potential of robotics was quite small and concentrated on the displacement question entirely. (Ayres and Miller, 1981a, pp. 134-135)

No primary data base exists from which to estimate the number of jobs that will be created by the robotics industry in the U.S. or Michigan. Normal U.S. government statistical sources are no help whatsoever since robotics is not a separately identified industry. According to the Michigan Employment Security Commission (MESC), the Bureau of Labor Statistics (BLS) instructions are to place firms that produce lifting and handling robots in SIC code 3537 (Industrial Trucks and Tractors), and firms that produce all other types of robots in SIC code 3569 (General Industrial Machinery, Not Elsewhere Classified). Given the small size of robot manufacturing today, such a classification scheme insures that no meaningful information about robot manufacturing will be available from government sources for some years.

Yet, there remains a significant need for such data, particularly occupational data. Interest in community college curricula for robotics technicians appears high nationwide, and formal course offerings are proliferating. True, there may be a shortage of trained technicians today (not to mention engineers with robot applications experience); but robotics is a "glamour" field, so there is also the possibility of turning a shortage of technicians into a glut in the near future. In any case, meaningful projections of employment by occupation in robotics are needed to guide potential new entrants to the labor force as well as those who may be seeking retraining.

Given the paucity of data, and other factors that will become apparent shortly, this chapter is necessarily more speculative than previous chapters. It is an initial effort to estimate the potential for job creation due to robotics in the U.S. and Michigan by industry and occupation. We will explain our forecast and assumptions sufficiently to enable (or encourage) those who follow to improve on our efforts.

Our estimates were developed from interviews, secondary sources, and where necessary, our own judgment. The data for the robot manufacturing occupational profile were provided under conditions of strict anonymity and with the understanding that only broad aggregates would be published. Although the occupational profile is only an informal representation of the robotics industry, it does represent the bulk of the output of that industry. Complete data for the occupational profile were obtained from firms representing well over 50 percent of the output of the industry.

The chapter is organized as follows. A brief description of the robot industry today is provided first. Then, the general methodology to estimate the employment impacts in 1990 is presented, including the limitations of that methodology. Third, the total 1990 employment impacts and the specific occupational impacts are discussed, for the U.S. and then for Michigan. The chapter concludes with an examination of the training implications of robotics.

Robot Manufacturing Employment

The U.S. robot manufacturing industry today is still embryonic, accounting for approximately $150 million of sales in 1981. Conigliaro's estimates of sales by firm for 1981 are provided in table 4-1. The firms with Michigan production facilities—Prab Robots, Inc., Copperweld Robotics, and DeVilbiss—accounted for 19 percent of the estimated sales. Clearly, however, the industry is dominated currently by Unimation in Danbury, Connecticut and Cincinnati Milacron, whose headquarters and research facilities are located in Ohio and whose production facilities are located in South Carolina. Conigliaro says that roughly 20 percent of U.S. production is exported (Conigliaro, June 19, 1981, p. 2) while the UM/SME Delphi forecast estimates that imports today are 20 percent of sales. Given the European and Japanese expertise in robotics technology, it is somewhat surprising that the UM/SME Delphi forecast expects the latter percentage to remain constant through 1990. (Smith and Wilson, p. 45)

Our interviews revealed considerable disappointment in 1982 sales, but that must be weighed against the optimistic (perhaps overly optimistic) sales expectations that prevailed earlier in the year, as discussed in chapter 2. Of course, there also has been some entry of new firms, so smaller-than-expected sales may in part reflect smaller market shares. There are reports, however, that the new entrants are not doing well in terms of orders. (''A Robotics Mecca in Michigan? Car Sales Must Rebound First'')

Table 4-1
Estimated Sales of U.S. Robot Manufacturers, 1980-1981*

Robot manufacturers	Sales 1980 (in millions)	1981
Unimation	$ 42	$ 55
Cincinnati Milacron	30	40
Prab Robots, Inc.	6	10
ASEA	7.5	12
Copperweld Robotics	4.5	6
Advanced Robotics Corp.	-	5
Automatix	0.4	3
Cybotech	-	3
Nordson Corp.	1	3
DeVilbiss	8	12
Mobot Corp.	0.7	1
U.S. Robot	-	0.8
Other	1	2
Total	$100**	$150**

SOURCE: Laura Conigliaro, *Robotics Newsletter,* No. 3 (New York, NY: Prudential-Bache Securities, Inc., March 25, 1981), p. 2.

*Conigliaro actually provides a range for sales. We show only the lower end estimates because actual total sales in both years were near the lower end of her range according to most sources.

**Total sales is itself an estimate, so the individual estimates of firm sales do not sum to the total sales estimates.

As stated in chapter 2, we expect sales revenues were flat in 1982 or perhaps showed a small increase. In any event, sales growth in 1982 was much less than that required to support some of the higher growth projections. Although prospects for 1983 are difficult to assess at this time, there appears no basis to expect near term improvement unless a vigorous economic recovery begins soon.

The consensus of economic forecasters is that general economic growth in 1983 will be modest at best. The U.S. auto companies, after experiencing the worst year since 1961 in the 1982 model year, remain cautious about production plans for 1983 models even though there was some sales improvement in late 1982. ("Motor Vehicles, Model Year 1982," p. 23) Economy-wide reports of 1983 capital spending plans, including machine tool orders, are especially sluggish. ("Business Outlays to Rise Modestly in '83 First Half"; "Little Corporate Zest for Leading a Recovery," p. 14; "Capital Spending's Sickening Fall," p. 36) In short, strong recovery in robot sales, at least in the first half of 1983, appears unlikely.

We estimate employment in U.S. robot manufacturing currently to be approximately 2,000 workers nationwide. Consistent with normal BLS practices, this estimate includes foreign firms with U.S. production facilities such as the Swedish firm, ASEA, but it excludes the sales and service offices of robot importers, even those with U.S. affiliates who serve as distributors. The BLS intent is simply to limit the definition of a particular manufacturing industry to actual domestic producers without regard to ownership. Given the rather rapid entry of new firms in this industry, our estimate of 2,000 workers is only a rough approximation of 1982 employment levels.

Our estimate of the current occupational profile of U.S. robot manufacturers is presented in table 4-2. For comparative purposes, the occupational structure of the motor vehicle and equipment industry, all manufacturing, and all industries are also presented. The employment profiles have been aggregated in the listed occupational groupings primarily to facilitate comparison and to highlight the technical labor input component. Unquestionably, the most surprising finding is that slightly over two-thirds of the workers in robot manufacturing are in the traditional white

Table 4-2
Current U.S. Occupational Profiles: Robot Manufacturing, Motor Vehicles and Equipment, All Manufacturing, and All Industries

Occupation	Employment distribution (percent)			
	Robot manufacturing	Motor vehicles & equipment	All manufacturing	All industries
Engineers	23.7	2.3	2.8	1.2
Engineering technicians	15.7	1.2	2.2	1.4
All other professional and technical workers	4.2	2.4	4.0	13.5
Managers, officials, proprietors	6.8	3.3	5.9	8.1
Sales workers	3.4	0.5	2.2	6.3
Clerical workers	13.9	6.2	11.3	19.9
Skilled craft and related workers	8.4	20.8	18.5	11.8
Semi-skilled metalworking operatives	4.2	15.8	7.2	1.7
Assemblers and all other operatives	19.0	38.6	36.2	13.1
Service workers	-	2.8	2.0	15.8
Laborers	0.7	6.1	7.7	6.0
Farmers and farm workers	-	-	-	1.0
Total	100.0	100.0	100.0	100.0

Columns may not add to total due to rounding.

collar areas of the professions, technicians, administrators, sales and clerical workers, while only one-third are in the traditional blue collar areas of the crafts, production operatives, and laborers. To some extent that is simply a reflection of a young, high technology industry with low sales, where the firms tend to be assemblers with little fabrication of parts. However, it is also indicative of a product that cannot be produced and sold like a loaf of bread; there are significant requirements for engineering design, programming and installation for *each* specific application.

Engineers are the dominant occupation in robot manufacturing (and a large number of the managers, officials and proprietors are trained engineers also). We estimate that 23.7 percent of the robot manufacturing employees currently are engineers. The bulk of these jobs are held by mechanical and electrical engineers, although there are a large number of electronic/computer specialists as well. There are also "proposal sales engineers" who prepare detailed plans and cost estimates based upon information from sales representatives. One manufacturer described "proposal sales engineers" as the heart of the business and claimed that only the best engineers are assigned the task. We did not classify these people as sales workers because it would not be indicative of the training required for the job. We estimate that no more than one-fourth of the engineers are working in research and development efforts at present.

The second most prevalent occupation, engineering technicians, represents 15.7 percent of the workforce. The bulk of these jobs could be called "robotics technicians," although there are also drafters, mechanical engineering technicians, and electrical and electronic technicians. The term robotics technician is more generic today than descriptive of a specific occupation with clearly defined training requirements. One manufacturer was not even aware of community college graduates in this field. It is likely that as two-year graduates become readily available, manufacturers will mold the job to the tasks for which the technician is best trained or for which an aptitude exists. Currently, the most prevalent tasks for robotics technicians in robot manufacturing are testing, programming, installing and troubleshooting, both in the manufacturer's plant and on-site with purchaser of the robot. Some robotics technicians also function as trainers and manual writers. One manufacturer speculated that perhaps some might become sales representatives. To some extent, robotics technicians are the key to ameliorating any possible shortage of trained robotics personnel in the short run. Robotics technicians are also needed for maintenance tasks by corporate users of robots, a topic which is discussed later.

Together, engineers and technicians constitute nearly 40 percent of all employment in robot manufacturing. That must be tempered with the knowledge that the industry is very small absolutely, so 40 percent of robot industry employment probably represents less than 1000 jobs nationwide. The future prospects for engineers and technicians in robot manufacturing are discussed later.

The concentration in the technical areas is offset by a relative lack of jobs in the production worker occupations typical of more conventional manufacturing industries. Table 4-2 shows a marked lack of craft workers, semi-skilled metalworking operatives, assemblers, and laborers when compared to other manufacturing. Clearly, this reflects the low level of robot production, but it also reveals the high technology component of robotics.

In general, robot manufacturing can be contrasted with other manufacturing by the rather obvious "skill-twist" of the occupations. Over two-thirds of the jobs are white collar versus much less than one-third in all manufacturing. Well over 50 percent of the jobs in robot manufacturing require two years or more of college training versus less than 20 percent in all manufacturing. Even assemblers in robot manufacturing generally perform higher-order assembly tasks than most assemblers in other manufacturing industries.

Robot-Related Employment

Besides employment in robot manufacturing itself, there are also numerous jobs created directly in other industries as a result of the spread of robotics technology. Robot-related employment exists today in firms that are direct suppliers to robot manufacturers and in firms that use robots. Some employment is also beginning to emerge in what we designate as robot systems engineering—primarily the installations or applications engineering required for robot systems. This area may or may not actually develop into an identifiable sector but it will likely create additional employment opportunities nonetheless.

Direct Suppliers to Manufacturers

Robot manufacturing directly creates jobs in firms that supply the parts and components (material inputs) to make a robot. Table 4-3 details the major components of a composite robot, the industry of origin by SIC code (3-digit level of aggregation), and the value of the material inputs supplied by each industry as a percent of the total value of material inputs. The information in the table was provided by William R. Tanner, a robotics expert and engineering consultant. The total value of material inputs makes no allowance for shop labor to assemble a robot, applications engineering, or any overhead costs. Naturally, these parts and components actually vary somewhat depending on the in-house capabilities of the robot manufacturer and the type of robot being produced.

Table 4-3
Major Component Parts of Robot by Industry of Origin

SIC code	Industry	Major parts of robot	Percent of total value of material inputs[a]
304	Rubber and plastics hose and belting	Pneumatic hose, rubber belting, V-belts	1
306	Fabricated rubber products, not elsewhere classified	Rubberized fabrics, grommets, tubing	
307	Miscellaneous plastics products	Vulcanized fiber, foams, molded plastic parts, custom compounds of resins	1
329	Abrasive, asbestos, and miscellaneous nonmetallic mineral products	Gaskets, grease seals, oil seals	< 1
331	Blast furnaces, steel works, and rolling and finishing mills	Steel pipes and tubes	1
332	Iron and steel foundries	Malleable iron castings	2
335	Rolling, drawing, and extruding of nonferrous metals	Copper wiring and tubing	< 1
336	Nonferrous foundries (castings)	Aluminum castings	1
339	Miscellaneous primary metal products	Heat treated metal parts	1
343	Heating equipment, except electric and warm air, and plumbing fixtures	Heat exchangers, radiators	1
344	Fabricated structural metal products	Manufactured sheet metal forms and machine guards	< 3

345	Screw machine products, and bolts, nuts, screws, rivets, and washers	Bolts, nuts, screws, rivets, washers	<1
346	Steel forgings and stampings	Electronic enclosures, perforated stamped metal	<3
349	Miscellaneous fabricated metal products	Valves and pipe fittings, wire springs, fabricated pipe and pipe fittings	1
356	General industrial machinery and equipment	Pumps, ball and roller bearings, blowers for exhaust fans, air filters, speed changers, gears, ball joints, clutches, couplings, drive chains, sprockets, pulleys, fluid power motors, fluid filter elements	17
357	Office, computing, and accounting machines	Electronic computing equipment	14
359	Miscellaneous machinery, except electrical	Cylinders, machined parts on job basis	30
362	Electrical industrial apparatus	Electric motors, synchros, electromagnetic brakes and clutches, electric motor controls and starters, positioning controls, solenoid switches, controls and control accessories	18
364	Electric lighting and wiring equipment	Current-carrying wiring devices, non-current-carrying wiring devices	1

Table 4-3 (continued)
Major Component Parts of Robot by Industry of Origin

SIC code	Industry	Major parts of robot	Percent of total value of material inputs[a]
367	Electronic components and accessories	Semiconductors and related devices, electronic capacitors, resistors, electronic coils, transformers, inductors, electronic connectors, printed circuits, switches	4
382	Measuring and controlling instruments	Industrial instruments for measurement, display and control of process variables, totalizing fluid meters and counting devices, instruments for measuring and testing of electricity, other measuring and control devices	<2

SOURCE: William R. Tanner, *Productivity Systems, Inc.*, Farmington, Michigan.

a. Total value of material inputs does not allow for shop labor to assemble robot, applications engineering, or any other overhead costs.

Table 4-3 is important because it indicates the direct supplier industries that will experience the greatest employment impacts due to the growth of robot manufacturing itself. According to these results, five industries account for the bulk (83 percent) of the value of all material inputs. The computer or microprocessor and other associated electronic hardware are provided primarily by the SIC code 357 (Office, Computing, and Accounting Machines) and SIC code 367 (Electronic Components and Accessories) sectors respectively. Together, these two industries account for approximately 18 percent of the value of material inputs. That percentage may appear low to those unfamiliar with robotics technology but today's robot does not require a complicated general purpose computer. SIC code 356 (General Industrial Machinery and Equipment) provides various pumps, motors, gears, speed changers, etc., and accounts for 17 percent of the value of material inputs. Electric motors and controls and other electrical apparatus is provided by SIC code 362 (Electrical Industrial Apparatus). These items account for approximately 18 percent of the value of material inputs. Finally, machine shops that provide precision-cut steel or steel alloy parts constitute the largest single proportion of the value of material inputs, approximately 30 percent. These machine shops are classified in SIC code 359 (Miscellaneous Machinery, Except Electrical).

As shown in table 4-3 there are numerous other industries involved in supplying component parts to robot manufacturers, but each of them is relatively minor. In total, these other industries provide about 17 percent of the material inputs for a typical robot. The listing of the major parts of a robot is long but the components themselves do not stretch the bounds of existing technology or the manufacturing capabilities of supplier firms. To some extent it is true that the robot itself represents old technology. The challenge is to extend robot capability and reliability while successfully integrating them into specific production processes.

Robot Systems Engineering

The process of integrating robots with other plant equipment is usually called installations or applications engineering. The bulk of the installations engineering today is being performed by robot manufacturers or by the purchasers themselves. There is no evidence yet, however, that in-house plant engineering staffs of user firms are being expanded to accommodate the introduction of robots, although some think that must (or should) be done. Robots today simply are an added responsibility for plant engineering staffs of user firms. Thus, the current situation raises few interesting robot-related employment questions. As robot systems become more numerous, however, there may be significant changes in applications engineering responsibilities.

According to the UM/SME Delphi forecast, 20 percent of industrial robots will be purchased as part of robot systems (versus individual stand-alone units) by 1985. That figure is expected to climb to 40 percent of all sales by 1990. (Smith and Wilson, p. 46) Even small robot installations of one or two units can be complicated, but larger installations of robot systems present many more predesign and technical integration problems. The robots must be interfaced not only with other plant equipment but also with each other; the details of planning and design expand geometrically. Additional applications engineering capability will be mandatory with such complex systems. The question is who will provide these applications engineering services.

In our interviews, considerable interest on the part of users was expressed in outside engineering assistance for robot installations. Some even indicated a desire for so-called "turn-key" robot systems. The term "turn-key" is applied (as in computer systems applications) to systems providers who are able to completely install one or more robots and all associated peripheral equipment, including any special purpose or hard automation. The "turn-key" provider guarantees operation of the system and turns it over to the purchaser only after successful operation, hence the name "turn-key." Some robot manufacturers, independent robot systems consultants (who are not robot manufacturers), and traditional machine tool or dedicated automation providers, have all indicated interest in the market for "turn-key" systems. A partial list of the firms who have either announced entry or who are expected to enter this market include: Unimation, Cincinnati Milacron, Bendix, Cross and Trecker, F. Joseph Lamb, General Electric, IBM and Texas Instruments. This market is attracting so much attention because the systems provider acts as a general contractor and therefore may come to be influential in total factory automation purchase decisions.

It is not necessary in this study to determine either who will provide robot systems or whether significant markets will emerge for "turn-key" robot systems. However, it is important to note that robot systems will require significant applications engineering capabilities that will likely add to robot-related employment. In part, the strong desire of users of robots for outside assistance in performing robot applications engineering is just another reflection of the lack of adequately trained personnel who truly understand the capabilities and limitations of robotics technology.

Besides applications engineering, installation of robot systems also requires considerable peripheral equipment and special purpose or dedicated machinery, usually denoted as the hard automation in the system. In general, such equipment has been provided by the traditional machine tool or capital goods sector. Thus one might plausibly argue that the hard automation in robot systems may create net new employment in the traditional machine tool sector; but that scenario appears to be of dubious validity, at least judging by the experience in the auto industry.

The bulk of all robot investments in the auto industry are being made during normal retooling or major overhauls of plant and equipment to produce new models. It does not appear logical to expect robot systems to create *additional* demands for the special purpose (and custom designed) machinery for the fabrication of auto parts. The composition of some of the support equipment like conveyors will certainly change but not necessarily substantially increase. In brief, the bulk of the hard automation in robot systems may be the identical machinery or slightly different machinery from what would have been used in the absence of robots. Economists call this capital for capital substitution. The true extent of the substitution is uncertain and beyond the scope of this study to determine empirically. However, it will likely be greatest in industries that are using significant amounts of automated equipment already, the same mass production industries most likely to adopt robots in the first place. Thus it appears doubtful that the installation of robot systems will be a significant plus for the traditional machine tool sector in terms of *net* new employment.

It should also be emphasized explicitly that the traditional machine tool firms may increasingly experience serious competition from the new providers of robot systems. The potential loss to the old-line firms includes not only the general contracting and design but also the possibility that at least some of the hard automation will not be sub-contracted to these firms either. In short, the introduction of robot systems may significantly alter long-standing relationships between firms and their traditional machine tool providers. No doubt, that is the very reason that some of the larger machine tool providers to the auto industry have themselves announced entry in the new market for robot systems.

Robot User Firms

The final area in which a significant number of jobs will be available is robot maintenance at corporate user locations. Typically, production up-time requirements are so high that maintenance must be available immediately in the case of robot failure. There are even stories of robots literally cut from the line and replaced with human workers to maintain production schedules. However, experienced personnel trained in complex machinery repair are not intimidated by robots and in fact are performing robot maintenance today with three months or less of training. We define these maintenance personnel as robotics technicians, but in firms with small numbers of robots such technicians will be required to maintain a variety of automated equipment. Even in larger firms, flexibility may be required of such technicians.

In the auto industry currently, robot maintenance technicians are primarily skilled electricians who have received specialized training in robot maintenance. Since robots are production equipment and all production equipment is maintained by members of the skilled trades bargaining unit of the UAW, these jobs will remain within the UAW in autos. As developed later in the chapter, it is doubtful that any of these jobs will be available to new graduates of two-year robotics technician programs in the near term.

At least two other job specialties are sometimes mentioned as potentially significant new robotics-related employment opportunities at corporate user firms: robot operators and robot programmers. We believe there is little potential for either. Presumably, robot operators would have minimal robotics training and oversee the operation of one or more robots. Such a function however, appears contradictory and self-defeating if the robot or robots were purchased to replace human workers (save labor costs). Provided sufficient electrical and/or mechanical limit or stop-switches were installed originally to properly interface the robot with other plant equipment, an operator should not be required in normal circumstances. In short, the term robot operator appears to be a misnomer, logically inconsistent and unlikely to emerge as a separate new occupation or employment opportunity.

The employment possibilities for robot programmers at corporate user locations are slightly more difficult to deal with. In nontechnical terms, today's robots are preprogrammed with a general software package that will enable the robot to accept (and remember) a specific routine. The specific routine itself is usually programmed and fully tested by the manufacturer on behalf of the purchaser of the robot. Once installed at the purchaser's location, today's industrial robots are usually not reprogrammed. However, to the extent that reprogramming is necessary for some specific applications, robot software packages, like other computer software packages, are made to be "user friendly." In our interviews, one robot manufacturer claimed that robot programming can be learned in two hours or less. That may be optimistic but certainly robotics technicians or others with similar skills can quickly learn to program robots with specific routines and in fact are doing so today as part of their regular duties.

Highly skilled computer specialists are required to develop the general software packages for robots, and more sophisticated robots and robot systems will increase the complexity of that software, particularly the requirements for interfacing the robots with plant equipment. However, as long as reprogramming tends to be infrequent or does not require changes in the general software, these positions will remain small in number and likely will continue to be found at robot manufacturers, specialty suppliers to robot manufacturers, or possibly with robot systems providers, rather than at user sites.

The purpose of the discussion of robot-related employment was to lay the necessary groundwork for our projections of job creation associated with the spread of robots in the U.S. and Michigan. Robotics-related jobs exist today in direct suppliers to robot manufacturers and in robotics maintenance at corporate user sites. As of now, the robotics applications engineering is being done largely by robot manufacturers and/or by the purchasers. However, the growth of robot systems will likely create new employment opportunities in this area. With these general categories of employment established, we are prepared to present our forecast of specific job creation accompanying the spread of robotics technology.

Forecast of Job Creation Due to Robotics

One of our goals in this study was to develop a consistent economic framework within which to estimate the impact of industrial robots in the U.S. and Michigan. To some extent, the specific methodology to forecast job creation is predetermined or conditioned by other parts of the study, although there are some unique issues in job creation. General methodological issues are discussed first; this includes the range of the estimates, the specific industries and/or areas in which jobs will be created, and the limitations of our approach. Then the individual industry forecasts are presented with an explicit discussion of any unique assumptions that apply to each. Finally, a summary occupational structure of the jobs created is presented.

As stated earlier, some of the methodological issues in job creation are predetermined or conditioned by other parts of the study. The projection date remains 1990. The range in the expected U.S. population of robots from 50,000 to 100,000 implies an annual robot sales level in 1990 of 14,175 to 28,350 robots, assuming a constant rate of growth throughout the forecast period. The exponential growth assumption is artificial, but no one can predict the peaks and valleys of the business cycle; so there really is no viable alternative to assuming a 1990 sales level which is consistent with the average growth needed to achieve the projected 1990 population of robots.

As mentioned earlier, Conigliaro estimates robot exports as 20 percent of production today, and the UM/SME Delphi study estimates imports at a constant 20 percent of sales through 1990. In the absence of any better information, we have assumed imports and exports will roughly offset each other. Consequently, a 50,000 to 100,000 range in the U.S. population of robots in 1990 is still consistent with U.S. production of 14,175 to 28,350 robots in 1990.

However, there is no guarantee that American producers will hold their share of the worldwide robot market. If exports were to fall or imports were to rise significantly, the employment effects would be correspondingly reduced in the manufacture of robots and in robot manufacturing suppliers. This threat is especially menacing because of Japanese and European expertise in robotics technology. It is important not to delude ourselves; just because the U.S. may be a large market for robots, there is no guarantee that those robots will be manufactured here.

There is also the question of robot replacement demand in 1990, although this is less difficult to deal with than the export/import market. Because the population of robots is so small today and because the lifespan of robots is expected to be a decade or more, there will be very little replacement demand until well into the 1990s. Even if there are significant breakthroughs in robotics technology in the near future, we do not expect obsolescence to become a factor in the demand for robots, since they are capital goods and can be expected to generate productive services for many years.

The industries and/or areas within which jobs will be created directly by industrial robots were introduced in our earlier discussion of the robotics industry today; namely, robot manufacturing, direct suppliers to robot manufacturers, robot systems engineering, and corporate users of robotics technology. Corporate robot users are again separated into autos and nonautos to maintain comparability with the job displacement figures in the previous chapter.

We will estimate the likely applications engineering requirement for robot systems without specifying the industry within which that employment will occur. It might be argued that corporate users will increase their engineering staffs to support the development of robot systems; but there is little evidence of that so far. It is possible that the robot manufacturers will best understand robot systems and therefore will sell their products inclusive of applications engineering services. It may be that machine tool builders who already act as general contractors for automated systems are best suited to provide the new robot systems. Finally, one could argue that independent robotics specialists (consultants) will evolve to supply robot systems. Our interviews supported all of these viewpoints and more. However, the truth is that no one knows the likely market outcome, so we attempt to estimate the applications engineering requirement without regard to industry of employment.

In summary, we estimate total employment in the U.S. in 1990 *directly* due to robotics in four broad areas: robot manufacturing, direct suppliers to robot manufacturers, robot systems engineering, and corporate robot users, the latter to identify maintenance requirements for robots. The projected employment impacts are based upon estimates of annual sales in 1990 that are consistent with the total population of robots forecast in chapter 2.

Although we believe that the above is the best possible procedure to estimate job creation given the constraints of this study, there are limitations and caveats which must be stated. First, we have not estimated the induced income effects that lead to further job creation. Clearly, the new employees in robotics spend their income which creates further jobs in wholesale and retail trade, services, etc. However, we have not estimated the *negative* induced income effect of jobs displaced either. Suffice it to say that the induced income effects, both positive and negative, raise complex issues that are beyond the scope of this study and neither is investigated here. Our method does include the total impact of purchases of material inputs by robot manufacturers, however.

A closely related limitation is that we do not consider price or competitive effects. As stated earlier, robots are introduced to lower costs and improve the quality of the product. If price reductions result, this will add to demand. Thus the productivity gains are passed along to the consumers of the product, and the increasing level of sales induces some "second-order" job creation. The other side of this argument is that without robots there would be additional job losses in those industries falling prey to foreign competition, so displacement questions can be ignored. The responsiveness of demand to price (called elasticity of demand by economists) and competitive effects are legitimate issues. We are not able to make specific forecasts of their magnitude, however, so these issues are not directly addressed here.

A final limitation is that we estimate total direct job creation rather than the increase in jobs from now to 1990. In other words, some jobs already exist today in robotics in the U.S., and those impacts have already been registered in the national economy. In chapter 3 we disregarded any intermediate displacement effects that had already occurred, so we do likewise in this chapter on job creation. In any event, the precision of our estimates is not sufficient that 2,000 employees make a significant difference. Again, we are trying to establish the general order of magnitude of job creation in robotics by 1990; that is all that is possible at this early date.

The potential cumulative direct job creation in the U.S. by 1990 due to robotics is presented in table 4-4. We estimate that the total number of jobs created will range from a low of about 32,000 to a high of about 64,000. The low range of the estimate assumes a 1990 impact which is consistent with a U.S. population of robots of 50,000, and production of 14,175 robots in 1990. The high range of the estimate assumes an impact consistent with a population of 100,000 robots in 1990 and production of 28,350 robots in 1990. Both the low and high estimates assume no change in the relative importance of the U.S. import or export market for robots.

Table 4-4
Direct Job Creation in U.S.
Due to Robotics, 1990

Area or industry	Employment: Range of estimate — Low	High
Robot manufacturing	8,700	17,400
Direct suppliers to robot manufacturers	8,091	16,182
Robot systems engineering	5,297	10,594
User firms - auto	3,000	5,000
User firms - other	7,000	15,000
Total	32,088	64,176

Robot Manufacturing

We estimate employment in robot manufacturing in the U.S. to range from a low of 8,700 to a high of 17,400 in 1990. This was estimated in the following way. Gross sales volume per employee by robot manufacturers exceeded $90,000 according to our interviews, while Conigliaro's estimated average price in 1981 slightly exceeded $70,000. (Conigliaro, June 19, 1981, p. 9) That implies an average output per employee of just under 1.3 robots.[1] Productivity was assumed to improve by a conservative 3.4 percent per year, the forecasted rate for all manufacturing contained in the National Planning Association's projections of the U.S. economy. (Terleckji and Holdrich, p. 4) So average output per worker in robot manufacturing would be slightly over 1.6 robots by 1990. Total U.S. employment in robot manufacturing in 1990 was then determined by dividing the potential market of 14,175 units to 28,350 units by average output per employee.

We believe that 1990 employment in robot manufacturing is likely overestimated using our procedure. First, 1982 has not been a good year for robot manufacturers, so sales per employee may not reflect average conditions in the industry. Also, there may have been overstaffing due to expected strong growth that did not materialize. Both these facts lead us to think that sales per employee, the numerator in estimated robot output per employee, is probably underestimated. Second, Conigliaro's average price (the denominator in estimating average robot output per employee) was a 1981 price, while prices are know to be falling in 1982. So the denominator is probably overestimated. In addition, the productivity improvement factor of 3.4 per-

1. Our rough estimate of employment in robot manufacturing of 2,000 included an estimate of employment in some firms with little or no robot sales in 1982. For obvious reasons, such firms were excluded from the calculations for average robot sales per employee and therefore average robot output per employee.

cent annually appears extremely modest for such a young industry. These industries are the very ones that sometimes exhibit spectacular productivity gains. All of these factors lead us to think that average robot output per worker is underestimated in our calculations. This has the effect of overestimating the employment in robot manufacturing consistent with a given level of sales. This approach was used nonetheless because of our decision to use known facts and conservative assumptions throughout the study.

Direct Suppliers to Manufacturers

We estimate employment in firms that are direct suppliers to robot manufacturers in the U.S. to be from about 8,000 to 16,000. We followed the approach of Burford and Katz where the direct supplier jobs or interindustry effects can be found as a multiple of the jobs in the primary industry, robotics in our case. (Burford and Katz, p. 152) The Burford-Katz technique can be applied in a nation or region (the latter will be helpful later in developing estimates for the State of Michigan) to estimate the direct supplier jobs where an input-output table which details the interindustry relationships is not available for the industry in question, or the industry itself is new so the interindustry relationships are unknown. Specifically,

$$M_j = \frac{w_j}{1 - \bar{w}}$$

where
M_j = relative effect of industry j on supplier industries in the region.

w_j = expenditures within the region of the jth industry as a proportion of total shipments.

$\bar{w}$ = average expenditures of all industries in the region as a proportion of total shipments.

Descriptively, the Burford-Katz technique is simply saying that the direct supplier effect in a nation or region is dependent on the uniqueness of the industry in question, which is measured by the purchases the industry makes from other industries as a proportion of its total shipments (w_j) and on the average interrelations among all industries that exist within the nation or region ($\bar{w}$). Burford and Katz tested their approach against several regions where input/output tables were available and found their short-cut method to be very precise. Of course, that kind of precision will not be achieved here, but it is possible to obtain a reasonable estimate of the direct supplier effect of robot manufacturing in the U.S.

Estimation of the direct supplier effect using the Burford-Katz approach is relatively straightforward. Step one was to estimate $\dot{w}$, essentially the material purchases of all firms in the nation as a proportion of total shipments. The total materials purchased by all U.S. manufacturing industries as a proportion of sales was 57 percent in 1977 according to the *1977 Census of Manufactures.* (U.S. Department of Commerce, 1981a, pp. 30-31) So a reasonable estimate of $\bar{w}$ is .57.

The second step was to estimate $\dot{w}_j$, the average material purchases of robot manufacturers in the U.S. Our interviews revealed a remarkable consistency in estimates of material purchases as a percent of sales in robot manufacturing—40 percent. So, if 40 percent of total sales of robot manufacturers are material purchases, and 57 percent of total sales of all manufacturing firms in the nation are material purchases, that leads to a direct supplier effect of .93. This implies that for every job created in robot manufacturing in the U.S., another .93 jobs are created in direct supplier industries. Thus, 8,700 to 17,400 jobs in robot manufacturing lead to 8,091 to 16,182 jobs in other manufacturing industries in the U.S.

In assessing the estimates of the direct supplier effect of robot manufacturing utilizing the Burford-Katz technique, several cautions and comments are worth mentioning. First, we do not account for import purchases of robot manufacturers or for all industries, the net effect of which may either lower or raise the direct supplier effect. If import purchases of materials are relatively more important for robot manufacturers than for all industries, then the direct supplier effect will be lower. Second, we do not allow for industry growth that sometimes increases interindustry dependence and the associated multiplier effect, especially for new industries like robot manufacturing. In both of these cases there is no empirical basis to determine the likelihood or magnitude of the possible change.

The third caution is that we utilize output relationships estimated with the Burford-Katz technique to determine employment impacts in the direct supplier industries. The resultant estimates will be true if and only if the supplier industries hire at the same rate as robot manufacturing. In this case that is probably acceptable since sales per employee in robot manufacturing is near the U.S. average for all manufacturing and the supplier effect itself is approximated by the U.S. average; but it remains a rough approximation only.

Given these cautions, we sought some confirmation of our estimate that (on average) for every job created in robot manufacturing .93 jobs are created in direct supplier industries.[2] That was done by examining related industries in the national input/output table for the U.S. where complete

2. Wassily Leontief, Institute for Economic Analysis, New York University, has undertaken a large project with support from the National Science Foundation to determine the impact of technological change on employment to the year 2000. The effort by Leontief holds the prospect of yielding more definitive information about the direct supplier effect of robot manufacturing and other emerging high technology industries.

interindustry relations are available, imports are accounted for, and employment to output relations are known directly. Unpublished data from the 151 sector national input/output table were provided by the Office of Economic Growth and Employment, Bureau of Labor Statistics. Specifically the 1981 total employment requirements table was utilized. This table shows total employment impacts per one million dollars of sales of the product of each industry to final users, based upon the interindustry relationships in the 1977 input/output table (the most recently available) and 1981 employment to output relationships.

As discussed earlier in the study, the most closely related industry to robot manufacturing is generally acknowledged to be numerically controlled machine tools; a more distant second might be computers. In the national input/output table, numerically controlled machine tools are a part of metalworking machinery. For that sector (SIC code 354) the 1981 employment requirements table indicates that on average, for every job created in metalworking machinery .73 jobs are created in supplier industries.[3] Similarly, for every job created in computers and peripheral equipment (SIC codes 3573-3574), 1.53 jobs are created in supplier industries.

When compared to our estimate of the direct supplier effect of .93 for robot manufacturing, the national input/output table indicates that the direct supplier effect of metalworking machinery is less than that for robot manufacturing, but the effect of computers and peripheral equipment is significantly greater. Of course, such comparisons are never clear-cut (metalworking machinery encompasses much more than numerically controlled machine tools) and do not constitute empirical proof in any event. Nonetheless, it is in-

3. The direct supplier effect was approximated as the total employment impact of the industry per million dollars of sales minus the impact in the primary industry itself with the result divided by the employment impact in the primary industry.

teresting that the national estimates of related industries do not contradict our estimates of the direct supplier effect for robot manufacturing; they even support the notion that robot manufacturing is more closely aligned with metalworking machinery, its most closely related industry in the national input/output table, than with computers.

Besides computers, we also calculated the direct supplier effect for the four other major parts and components suppliers for robots. The results are .71 for SIC code 359 (Miscellaneous Machinery, Except Electrical), 1.05 for SIC code 362 (Electrical Industrial Apparatus), .90 for SIC code 367 (Electronic Components and Accessories), and 1.19 for SIC code 356 (General Industrial Machinery and Equipment). If the direct employment impact of robot manufacturing is assumed to be the weighted average of these five industries which supply 83 percent of the value of material inputs to robot manufacturers, where the weights are the relative percentages from table 4-3, the direct supplier effect is 1.03, once again close to the impact estimated with the Burford-Katz technique.

In short, we conclude that the order of magnitude of our estimate of the direct supplier effect of robot manufacturing is very reasonable. The direct supplier effects of robot manufacturing may appear modest to some observers, but any industry where material purchases are only 40 percent of sales will likely have a small relative effect on other industries.

Robot Systems Engineering

Total employment in robot systems engineering in the U.S. was estimated as 5,300 to 10,600 in 1990. This is a net figure that represents the likely applications engineering employment, without specifying the actual industry of employment. As discussed earlier, that means these jobs may be with corporate users, robot manufacturers, machine toolmakers, or independent robot systems consultants. Since robot systems are not a separate market at this point, both total employment and the occupational profile are based on very sketchy information. First, total system sales in 1990 are estimated. According to the UM/SME Delphi study, the total robot market in 1990, including total system sales and individual unit sales of robots, will be nearly $2 billion in terms of 1980 dollars. (Smith and Wilson, p. 50) The UM/SME Delphi forecast also estimates that the 1990 average price for a robot will be $30,000 in terms of 1980 dollars. In addition, 40 percent of all robot sales will be system sales, and 30 percent of the cost of a system is the robots themselves. (Smith and Wilson, pp. 37, 46 and 47) That implies that the nearly $2 billion robot market in 1990 consists of $.6 billion in individual units, $.4 billion packaged for systems, and $.93 billion in other systems hardware. So total system sales in 1990 in terms of 1980 dollars would be $1.33 billion.

The UM/SME Delphi estimate of the total systems market in 1990 must be adjusted downward to reflect our smaller forecast of unit sales of robots in 1990. We estimate 1990 robot sales at 28,350 (maximum) while the UM/SME Delphi forecast implies unit sales of 33,333 ($1 billion divided by the average price of $30,000). Using the ratio of our estimate of unit sales to the UM/SME Delphi estimate of unit sales, our derived estimate of total systems sales of $1.13 billion in 1990 emerges.

The second step is to estimate the applications engineering required for system sales of $1.13 billion. According to our interviews, applications engineering constitutes about 30 percent of the cost of a robot system, so the applications engineering required for $1.13 billion total system sales would be approximately $340 million. We then assumed that approximately 80 percent of the applications engineering requirement is personnel costs, which results in $272 million wage income for applications engineering in robot systems. The third step is to estimate employment, given that the applications engineering represents approximately $272 million (1980 dollars) of wage income in 1990. That requires estimation of the occupational content of the jobs as well as average pay for those jobs. Those estimates are presented in table 4-5.

The occupational distribution for the applications engineering of robot systems is unknown today, so a hypothetical distribution was constructed based upon the occupational profile of robot manufacturers. The relative importance of (1) all other professional and technical workers, (2) managers, officials, and proprietors, (3) sales workers, and (4) clerical workers is the same as that for robot manufacturers. The remainder of the occupational profile, engineers and robotics technicians, assumes a two-to-one ratio between these occupations. That is based on the occupational profile of robot manufacturers who limit robotics technicians to testing, programming, troubleshooting, and installation of robots. Given that our estimates of the dollar value of the applications engineering of robot systems is stated in terms of 1980 dollars, the relative distribution of occupations can be used in conjunction with estimates of average earnings for these occupations in 1980 to arrive at total employment for robot systems engineering of 10,594 as illustrated in table 4-5. The identical procedure was followed for the low growth scenario which assumes a smaller market for robots in 1990. Although the details of those calculations are not discussed here, the net result is a minimum estimate of employment in robot systems engineering of 5,297.

Of course the separate estimation of approximately 5,300 to 10,600 jobs in robot systems engineering in 1990 and the occupational content of those jobs is highly speculative. This market barely exists today, and the future structure of the

Table 4-5
Robot Systems Engineering Employment in 1990

Occupation	Employment distribution (percent)	Total employment	Annual average pay in 1980 (dollars)[a]	Total pay (millions)
Engineers	47.8	5,064	29,806	150.9
Robotics technicians	23.9	2,532	19,896	50.4
All other professional and technical workers	4.2	445	24,984	11.1
Managers, officials, proprietors	6.8	720	32,461	23.4
Sales workers	3.4	360	27,253	9.8
Clerical workers	13.9	1,473	17,993	26.5
Total	100.0	10,594	—	272.1

a. Source of average annual pay data is U.S. Department of Commerce, Bureau of the Census, *Money Income of Households, Families, and Persons in the United States: 1980,* Series P-60, No. 132, p. 195.

market is unknown. At one extreme one might argue that we are guilty of double-counting, in that employment in robot manufacturing of 8,700 to 17,400 sufficiently accounts for the applications engineering requirement of robot systems. However, we think that future development of robot systems will cause a significant increase in the need for applications engineering to successfully install those systems, even though we cannot at this time identify the specific industry within which that employment will occur. At the other extreme one might argue that we are guilty of under-counting the jobs that will be created to provide the hard automation in robot systems. However, as discussed earlier, we think much of the hard automation in robot systems will be capital for capital substitution. Therefore the traditional machine tool sector will not experience significant net expansion due to the introduction of robot systems, although there certainly may be changes in the composition of the hard automation required and in the specific firms which provide it.

Robot User Firms

The estimates of the jobs created in the corporate user industries in the U.S. are 3,000 to 5,000 in autos and 7,000 to 15,000 in all other manufacturing. All of these jobs are assigned to robot maintenance, loosely called robotics technicians in this study, although such technicians may be required to maintain other automated equipment as well. Introduction of robot systems may require additional engineering support at corporate user firms as well, but those jobs have already been accounted for in our estimates of robot systems engineering employment.

The specific requirements for robotics technicians at corporate user firms were derived from Tanner and Adolfson's estimate that the maintenance standard is one person to ten robots per shift. (Tanner and Adolfson, p. 107) In some cases it is a low estimate, especially where total robot usage is low, but it is likely more accurate for more substantial installations and those installations should predominate by 1990. By that year, robot dependability should have improved significantly, but prudence dictates a conservative estimate. Using our assumption of an average of two shifts, two maintenance workers are needed per ten robots and the estimate of robotics technicians for maintenance of robots at corporate user industries follows directly.

Occupational Distribution of New Jobs

The occupational distribution of the jobs created by robotics in the U.S. is presented in table 4-6. The current robot manufacturing occupational profile was used in developing the 1990 profile, except that engineers were reduced to 20 percent of the total and assemblers were increased to 23.7 percent of the total. That corresponds to our expectation that engineers, especially in research and development, will not expand as rapidly as output, but assemblers will become relatively more important as output increases. The occupational profile for the direct suppliers to robot manufacturers was constructed as the weighted average of the individual occupational profiles for the five industries which account for 83 percent of the value of material inputs to a robot. The weights were determined based on the percentages in table 4-3. The individual occupational profiles (SIC codes 356, 357, 359, 362 and 367) themselves, are from the OES data base for 1980 and were provided by the Office of Economic Growth and Employment, Bureau of Labor Statistics. The robot systems engineering estimates utilized the occupational profile discussed earlier in this chapter. Finally, corporate user positions were all assigned to robotics technicians.

We estimate 4,600 to 9,300 engineering jobs will be created in the U.S. due to robotics and 13,000 to 26,000 engineering technicians will be required. The bulk of these technicians will be robotics technicians, but if one assumes that none of the jobs in autos will be available to two-year graduates of robotics technician curricula, then that figure should be reduced to 10,000 to 21,000, a rather significant difference. The training/retraining implications of these estimates are discussed later in this chapter.

Table 4-6
Direct Job Creation in U.S.
Due to Robotics, by Occupation, 1990

Occupation	Employment: Range of estimate Low	Employment: Range of estimate High
Engineers	4,636	9,272
Robotics technicians	12,284	24,568
Other engineering technicians	664	1,328
All other professional and technical workers	936	1,871
Managers, officials, proprietors	1,583	3,166
Sales workers	581	1,162
Clerical workers	2,908	5,817
Skilled craft and related workers	2,163	4,326
Semi-skilled metalworking operatives	2,153	4,306
Assemblers and all other operatives	3,763	7,526
Service workers	138	276
Laborers	279	558
Total	32,088	64,176

The other potential job creation in table 4-6 directly due to robotics does not appear to raise particularly significant training issues. The numbers are rather small and well within current supply capacity. It is important to note, however, that there is an extremely poor job match between the jobs that robots will likely displace and similar jobs that will be created through the introduction of robots. Specifically, in chapter 3 we estimated total job displacement of 100,000 to 200,000, with all of these tasks in the traditional blue collar areas, while from table 4-6 only 8,500 to 17,000 similar blue collar jobs will be created. Thus very few workers can expect to transfer to the new robotics sector but continue to perform essentially their old job tasks.

The shocking feature of table 4-6, the occupational profile of those jobs created, is that well over half of all of these jobs require two or more years of college training. That is consistent with the occupational profile of the robot manufacturing industry, but it is still a startling revelation and attests to the high technology nature of robotics.

Forecast of Job Creation in Michigan Due to Robotics

The job creation potential of robotics in the State of Michigan follows logically from the U.S. estimates. For that reason the organization of this section parallels that of the previous one. Additional assumptions and methodology are discussed as needed.

The level of the U.S. production of robots in 1990 alone is not sufficient to estimate the number of jobs created in Michigan, for we must also determine Michigan's market share of this production. Just as a single point estimate of the population of robots in 1990 was deemed unwise, it is also unwise to consider a single point estimate for Michigan's market share of that production. Since it is beyond the scope of this study to do a thorough locational analysis of the robotics industry, it is assumed that Michigan's share of the U.S. market in 1990 will range from a low of 20 percent to a high of 40 percent.

Both figures are speculative, but the low end of the estimate is conditioned by Michigan's current market share of the production of robots, approximately 19 percent in

1981. At first glance the high end of the estimate may appear overly optimistic, but it is not necessarily so. First, robotics is the number one (immediate) "target industry" in Governor Milliken's plan to attract high technology industry to the State of Michigan. (Milliken, 1981b, p. 13) Second, the proposed program of the nonprofit Industrial Technology Institute should help to attract robot manufacturers as well as other manufacturing process suppliers. Third, there is no doubt that the auto industry centered in Michigan is the single largest market for robots today. Fourth, some market entrants who have announced plans to produce robots or are producing robots already have a Michigan base. These include Bendix, General Motors, United Technologies, and a number of other small firms in the state.

There are no guarantees that these factors will increase Michigan's market share in the future. There are also no guarantees that imports or other factors will not reduce Michigan robot production below the low end estimate. While a market share of 20 to 40 percent for Michigan is optimistic, it is not unreasonable.

The potential cumulative direct job creation in Michigan by 1990 due to robotics is presented in table 4-7. We estimate that the total number of jobs created will range from a low of about 5,000 to a high of about 18,000. The low range of the estimate assumes a 1990 impact which is consistent with a U.S. population of robots of 50,000, production of 14,175 robots in 1990, and a 20 percent share of the market for Michigan. The high range of the estimate assumes an impact consistent with a population of 100,000 robots in 1990, production of 28,350 robots in 1990, and a 40 percent share of that production for Michigan. The range of the estimate for job creation is large because of the dual uncertainties of growth in the U.S. production of robots and the relative share of that production accounted for by Michigan.

Table 4-7
Direct Job Creation in Michigan
Due to Robotics, 1990

	Employment	
	Range of estimate	
Area or industry	**Low**	**High**
Robot manufacturing	1,740	6,960
Direct suppliers to robot manufacturers	974	3,898
Robot systems engineering	1,059	4,238
User firms - auto	1,065	1,776
User firms - other	287	865
Total	5,125	17,737

We estimate employment in robot manufacturing in Michigan to range from about 1,700 to 7,000 in 1990. That was estimated in the following way. Michigan's share of robot manufacturing employment was found by multiplying the low estimate of 8,700 employees nationwide by the low estimate of Michigan's share of the production for that market (20 percent). The same was done for the high production estimate and the high estimate of Michigan's share of that production. That leads directly to the final result.

We project employment in firms that are direct suppliers of robot manufacturers in Michigan to be about 1,000 to 4,000. Once again, we followed the approach of Burford and Katz to determine the direct supplier effect. The estimates are more difficult and tentative here because utilization of the Burford-Katz technique requires knowledge of material purchases of all industries and robot manufacturing *within* the state. That information is not available, but it is possible to obtain an upper bound for the direct supplier effect of robots in the State of Michigan.

Step one was to estimate $\overline{w}$ for Michigan, essentially the material purchases of firms within the state as a proportion of total shipments. The total materials purchased by all Michigan industries as a proportion of sales was 61 percent in 1977 according to the *1977 Census of Manufactures* (U.S. Department of Commerce, 1981a, p. 101), so an upper limit of $\overline{w}$ is .61. Burford and Katz, however, suggest that seldom do the material purchases within a region exceed .40 as a proportion of total shipments. (Burford and Katz, p. 158) We assumed $\overline{w}$ equals .50 in Michigan, which, if true, means that over 80 percent of material purchases of Michigan manufacturing firms are made from other manufacturing firms in the state.

The second step was to estimate w_j for Michigan, the average material purchases of robot manufacturers in the state from other industries in the state. Specifically, the estimation of w_j for a region depends not on the total material purchases of all firms in the industry but on the regional proportion of material purchases of such firms in the region. We assumed that total materials purchased by robot manufacturers in Michigan as a proportion of total shipments approximated the national average of 40 percent and that by 1990 a maximum of 75 percent of the material purchases of robot manufacturers in Michigan would be spent with the state. So, if 40 percent of total sales of robot manufacturers are material purchases, and for those manufacturers in the State of Michigan 75 percent of that figure is spent within the state, then w_j equals .28, or 28 percent.

Given our estimates of $\overline{w}$ and w_j for Michigan, we can use the Burford and Katz formula from the previous section. That leads to a supplier effect of .56 which implies that for every job created in robot manufacturing in Michigan another .56 jobs are created in direct supplier industries.

Thus, the approximately 1,700 to 7,000 jobs in robot manufacturing lead to about 1,000 to 4,000 jobs in other manufacturing industries in Michigan.

Although the estimates of the direct supplier jobs in Michigan may appear small, we believe these estimates are reasonable or perhaps overestimated. First, the average material purchases of all Michigan firms within the region is almost certainly less than the 50 percent used in our calculations, which would lower the direct supplier effect. Second, the average material purchases for robot manufacturing firms in Michigan from other firms in the state is much less than 28 percent of total shipments today (75 percent of total material purchases) and likely in 1990 as well. The State of Michigan is not a significant producer of microprocessors and the other electronic/computer-related components of a robot. Furthermore, very few industries in any region purchase as much as 75 percent of their material inputs locally from firms that are actually local producers (not wholesalers). In general, the direct supplier effect of an industry in an open region tends to be less than the direct supplier effect for that same industry in the nation.

Total employment in robot systems engineering in Michigan was estimated as approximately 1,000 to 4,200 in 1990. Once again, this is a net figure that represents the applications engineering employment likely in Michigan without specifying the actual industry of employment. It was found in exactly the same way as robot manufacturing employment in Michigan. Specifically, Michigan's share of robot systems engineering employment is the national employment in robot systems engineering (5,297 to 10,594), multiplied by the associated share of robot production in the state (20 to 40 percent). The implied assumption is that the hypothesized share of robot production in the state is also applicable for robot systems engineering. That may or may not be true, but there appears to be no better alternative at this time.

The stakes are not small in robot systems for Michigan's auto-dominated economy where employment in the Detroit-based machine tool sector accounted for over 50,000 jobs in 1977. (Institute of Science and Technology, p. 67) That sector, of course, has primarily supported the auto industry. Thus, Michigan's traditional machine tool providers and other general contractors in the state may experience serious competition from firms outside the state to provide robot systems to the auto industry. The potential loss includes not only the general contracting and design, which is considerable in a robot system, but also the possibility that at least some of the hard automation will not be provided by Detroit-based capital goods firms either.

Of course, this pessimistic scenario is only one possibility, and Michigan's success in developing the expertise for robot systems engineering may also serve to retain jobs in the traditional machine tool sector in the state. Robot systems engineering is an important area where the Industrial Technology Institute may come to play a role. The program of the Institute is to include an Applications Consultant Program which will aid with specific automation application problems. The Institute also plans to initiate a continuing program of research on the integration of manufacturing automation which will be essential to full implementation of the flexible automated manufacturing concept. It is entirely possible that Michigan could develop a manufacturing systems design capability that would lead to significant export of both goods and services to other states.

The estimates of the jobs created in the corporate user industries in Michigan are about 1,100 to 1,800 in autos and 300 to 900 in all other manufacturing. These numbers follow directly from the projection of the robot population in

Michigan and the assumption that one robotics technician is needed per ten robots per shift for maintenance work. In sharp contrast to the national estimates, almost 75 percent of the robotics technicians in corporate user industries in Michigan will likely be in the auto industry. This occurs because the auto industry will likely continue to be the largest single user of robots and because of the relative concentration of the auto industry in the State of Michigan.

The total occupational impact of the jobs created in Michigan is presented in table 4-8. This table was constructed in exactly the same way as the national estimates. We estimate about 900 to 3,600 engineering jobs will be created in Michigan due to robotics and 1,900 to 4,900 engineering technicians will be required. The bulk of these technicians will be robotics technicians, but if one assumes that none of the jobs in autos will be available to two-year graduates of robotics technician curricula, then that figure should be reduced to 750 to 2,700, a rather significant difference for the state. The training/retraining implications of these estimates are discussed in the next section.

Training Implications

On the whole, these job creation numbers are rather modest. It might even be assumed that there are no serious training questions arising from the creation of less than 65,000 jobs over an 8-year period; however, there are a few areas that should be mentioned as possible problems. The first is engineers; electrical and mechanical engineers will be required in significant numbers if the industrial robot population is to grow as we project. There will also be needs for industrial engineers and computer specialists as well.

Engineers. As described repeatedly throughout the study, industrial robots do not just come ''off-the-shelf'' and onto the factory floor fully functional from the time they are

Table 4-8
Direct Job Creation in Michigan
Due to Robotics, by Occupation, 1990

Occupation	Employment: Range of estimate Low	Employment: Range of estimate High
Engineers	898	3,593
Robotics technicians	1,810	4,469
Other engineering technicians	108	430
All other professional and technical workers	159	638
Managers, officials, proprietors	266	1,065
Sales workers	108	432
Clerical workers	505	2,020
Skilled craft and related workers	318	1,275
Semi-skilled metalworking operatives	288	1,154
Assemblers and all other operatives	610	2,441
Service workers	17	66
Laborers	38	154
Total	5,125	17,737

plugged in. A significant number of graduate engineers will be required to help robots find their place in U.S. factories. Specifically, we forecast a need for 4,600 to 9,300 engineers, primarily for robot system design and implementation as well as for design work with robot manufacturers.

What are the prospects of obtaining the required engineering talent to support the development of robotics? Clearly, the answer to this question can only be obtained by looking at the total market for engineers rather than focusing on specific industries within that market. First, we will examine the likelihood of a sufficient supply of engineers. Then we assess the prospects for the overall demand for engineers. In general, this supply-demand approach leads us to think that there is no reason to be optimistic that sufficient engineers will be available in the decade of the 1980s to support all of the requirements for engineers.

Table 4-9 shows the experience with the supply of engineers at the bachelor's, master's, and doctorate levels for the period 1960 to 1980. The absolute numbers of both bachelor's and master's degrees in engineering have increased, although the proportion of all U.S. bachelor's and master's degrees that are awarded in engineering declined from nearly 10 percent in 1960 to 6 percent in 1980. From 1970 to 1980 alone there has been an absolute decline in the number of doctorates awarded in engineering while the number of master's degrees awarded has increased only slightly.

Labor market analysts are well aware of the volatility in engineering enrollments. Post-World War II production of engineers has had at least three distinct cycles. First, enrollments increased explosively 1946-50 as veterans returned in large numbers to the campuses with the educational aid available under the GI bill. That was followed by a precipitous decline in enrollments of over 50 percent by 1955 as these same benefits were terminated. The second cycle began in the late 1950s due to the Soviet launching of "Sputnik" and the subsequent American response which included various kinds of student aid and research support in the sciences and engineering. The student aid was authorized as part of the National Defense Education Act of 1958. Thereafter there was a long sustained rise in engineering enrollments which peaked 15 years later in 1973, although in absolute terms enrollments never exceeded the level of the early 1950s. Finally, after a decline in enrollments through 1976, the third and current cycle began when engineering enrollments began to rise once again.

Table 4-9
Engineering Degrees Conferred in U.S.

Year	B.S.		M.S.		Ph.D.	
	Engineering degrees	Percent of all degrees	Engineering degrees	Percent of all degrees	Engineering degrees	Percent of all degrees
1980	59,240	5.9	16,846	5.6	2,519	7.7
1979	53,720	5.4	16,193	5.4	2,517	7.7
1978	47,411	4.8	17,015	5.4	2,442	7.6
1977	41,581	4.2	16,889	5.3	2,599	7.8
1976	39,114	3.9	16,170	5.2	2,835	8.3
1975	40,065	4.1	15,434	5.3	3,151	9.2
1974	43,530	4.3	15,393	5.5	3,336	9.9
1973	46,989	4.8	16,758	6.3	3,560	10.2
1972	46,003	4.9	16,802	6.6	3,704	11.1
1971	45,387	5.1	16,347	7.1	3,654	11.4
1970	44,772	5.4	15,597	7.4	3,681	12.3
1969	41,553	5.4	15,243	7.8	3,377	12.9
1968	37,614	5.6	15,188	8.6	2,932	12.7
1967	36,188	6.1	13,885	8.8	2,614	12.7
1966	35,815	6.4	13,678	9.7	2,304	12.6
1965	36,795	6.8	12,056	10.7	2,124	12.9
1964	35,226	7.0	10,827	10.7	1,693	11.7
1963	33,458	7.4	9,635	10.5	1,378	10.7
1962	34,735	8.3	8,909	10.5	1,207	10.4
1961	35,866	8.9	8,178	10.4	943	8.9
1960	37,808	9.6	7,159	9.6	786	8.0

SOURCE: National Science Foundation, *Science and Engineering Degrees: 1950-80,* NSF 82-307, Washington, DC, 1982, pp. 21-32.

Because of this record there is very little reason to be confident that recent increases in engineering enrollments will be maintained in the decade of the 1980s. (Freeman, pp. 111-117) In addition, the total number of science and engineering degrees awarded has been falling since the mid-seventies, so recent gains in engineering degrees appear to have come at the expense of other scientific fields. There is also no evidence that the proportion of people in the labor force with engineering degrees is increasing. (National Science Foundation, 1982a, pp. 60-62) Finally, the unemployment rate for engineers is extremely low, 1.5 percent in 1980, and the National Science Foundation reports that few engineers are working outside their professional specialties involuntarily. (National Science Foundation, 1981c, pp. 15-16) So we cannot expect the supply of engineers to expand much in the decade of the 1980s through reabsorption of unemployed engineers or the reentry of engineers who are currently working in nonengineering jobs.

A closely related question about the supply of engineers is the adequacy of the training received. Specifically, are engineers prepared in such a way as to facilitate the introduction of new technologies such as robotics? An adequate answer to this question is far beyond the scope of this study, but some comments are offered briefly. First, as mentioned in chapter 2, our interviews did reveal some criticisms of today's engineers—particularly that they are overspecialized. That sentiment is echoed in an article by Gail Martin which discusses manufacturing engineering as a much needed multidisciplinary engineering specialty. (Martin, 1982b, pp. 22-26) Second, the National Science Foundation reports that 10 percent of faculty positions in engineering programs were vacant in the Fall of 1980. (National Science Foundation, 1981a, p. 1) Moreover, approximately 90 percent of all engineering colleges reported that in the last five years, staffing had become more difficult because of their increasing inability to match private industry salaries. The schools responded to this shortfall of faculty by both cancelling classes and increasing teaching loads.

Finally, the National Science Foundation also reports that the proportion of science and engineering faculty with recent doctorates (within the last seven years) has fallen to approximately 20 percent in 1980 from almost 40 percent in 1968. (National Science Foundation, 1981b, p. 1) That decline is typical of all the engineering fields. It indicates that our academic teaching and research staffs in engineering schools are not receiving the infusions of young talent generally believed necessary to remain vigorous. From table 4-9 it is clear that we are training fewer Ph.D.'s in engineering than even five years ago at the very same time that engineering enrollments are increasing and the proportion of engineering faculty with recent doctorates is reaching new lows. These divergent trends are largely explained by the strong market demand for engineers, but they are very disturbing nonetheless. We cannot continue to borrow from the future human resource pool indefinitely for the sake of immediate needs. We fear that the supply of engineering graduates may prove to be a limiting factor in the spread of robotics technology in the U.S.

Turning our attention to the demand for engineers, it is clear first of all that there is no current surplus of engineers. Unlike the market for most other college graduates, the market for engineering graduates remains tight. That is attested to by the low unemployment rate for engineers, but there are even better indicators of the short-run labor market tightness for engineers. According to a National Science Foundation survey of the recruiting experience of private industries, there is a definite shortage of electrical engineers. Employers were only able to achieve 41 percent of their hiring goals for electrical engineers in 1981. (National Science

Foundation, 1982b, p. 4) Only computer engineering personnel were in a tighter supply situation. The same publication reports that the supply of industrial engineers and mechanical engineers were roughly in balance with demand. The signals were somewhat confusing for industrial engineers, however; they could be in current shortage also. Employers who reported shortages of engineers attributed the problem to the growth of needs in their industry.

The longer run prospects for engineers can be assessed by examining the occupational projections of the BLS, using the OES survey data base. The BLS projected change to 1990 in employment for engineering personnel as a percent of the 1980 employment base is shown in table 4-10. Once again, a range is provided reflecting the low- and high-growth scenarios for the national economy. It may be helpful to keep in mind that the total labor force is expected to grow in the decade of the 1980s, so it may be more meaningful to discuss the growth of engineers relative to all occupations.

Table 4-10
Projected Change in Employment for Engineering Personnel 1980-1990

	Percent change 1980 - 1990			
	Total manufacturing		Total employment	
Occupation	Low	High	Low	High
Electrical engineers	35	47	35	47
Industrial engineers	29	42	26	37
Mechanical engineers	34	49	29	41
Computer specialists	48	62	58	70
All engineers	30	42	28	38
All occupations	15	24	17	25

SOURCE: Based on data from OES survey provided by the Office of Economic Growth and Employment Projections, Bureau of Labor Statistics, U.S. Department of Labor, Washington, DC.

The relative increase projected for all engineers ranges from 50 to 100 percent greater than that for all occupations depending on the specific forecast. In relative terms, the greatest projected increase is for electrical engineers and computer specialists, although the needs for industrial and mechanical engineers are well above average as well. The specialties projected to have the greatest relative increase are also the largest specialties today in absolute terms. So the order of magnitude in relative terms is identical to the order of magnitude in absolute terms for these four fields—computer specialists, electrical engineers, mechanical engineers, and industrial engineers. In general, in the longer run, the need for engineers is projected to grow at a much faster rate than that for all occupations.

There is also anecdotal evidence which seems to support an increased need for engineers. More engineers will be required for the coming defense buildup that may not be reflected in the data today. Moreover, recent federal budget trends suggest a shift in demand away from teachers and social workers and toward engineers and scientists. It can also be anticipated that any acceleration in the rate of capital investment or plant modernization in U.S. industry will lead to additional requirements for engineers. If America is going to be reindustrialized, we will require the assistance of a great many engineers. So there may be a significant increase in the demand for engineers in the U.S. in the decade of the 1980s, perhaps even more than currently anticipated.

Viewed in isolation, the need for 4,600 to 9,300 engineers to support the growth of robotics technology appears inconsequential. The high range of the estimate for total engineers required for all robotics applications by 1990 represents less than one-fifth of one full year's graduates utilizing the 1980 data as shown in table 4-9. This hardly appears to be an insurmountable goal, but some concerns arise when we look at the broader supply-demand conditions for engineers within which robotics must compete for personnel. On the supply side, engineering degrees conferred have been increasing recently, but the historical precedent of extreme volatility in engineering enrollment leaves doubt that the recent increases will be maintained throughout the decade of the 1980s. There have also been no increases in degrees conferred at the master's and doctorate levels, and colleges report that they are finding it more difficult to retain qualified faculty, particularly younger faculty. So, among the many other supply issues, there are serious questions about the ability of our colleges to continue to increase engineering enrollments without possibly compromising the quality of engineering education.

On the demand side we begin the decade of the '80s with a short-run deficit of engineers, and longer run projections anticipate an increasing demand for engineers relative to all occupations. Thus, any decline in enrollments or further increases in demand (which appear likely) would have the potential of creating a severe shortage of engineers in the decade of the 1980s. Even if robotics employers attract more than their fair share of engineering talent, a tight engineering labor market may impede the growth of robotics technology in this decade.

Robotics Technicians. The largest major occupational demand category identified in table 4-6 is that of robotics technicians. We expect from 12,300 to 24,600 such jobs will be created in the U.S. by 1990. As explained earlier, we are using robotics technician as a generic term to refer to the individuals who have sufficient familiarity with robotics technology to be capable of testing, programming, installing, troubleshooting and maintaining industrial robots. In addition, we have included robot maintenance and repair in user industries, although this is somewhat arguable based on the auto experience discussed below. In a supply sense, we expect this emerging occupation will be dominated by graduates of 2-year community college programs. Again, this may be less true of the maintenance and repair function in user industries, especially in the auto industry.

The auto industry's demands for personnel to maintain and repair their robots must be treated separately, because it appears that this demand will not be expressed in the external labor market. Judging by the plans at General Motors, these positions will be staffed by existing employees. To begin with, the labor agreement between the UAW and General Motors assigns responsibility for maintaining production equipment to the Skilled Trades Council. With the introduction of welding robots at GM, a memorandum of agreement was signed on March 15, 1972 outlining the specific work assignments relating to the "Welding Equipment Maintenance and Repair" classification. This document also sets out the training requirements for an apprentice program for welding equipment maintenance and repair. (Agreement Between General Motors Corporation and the UAW, pp. 176-185) In brief, the rules have already been negotiated, and the UAW Skilled Trades Council has jurisdiction over the jobs.

Secondly, the newly endorsed "Statement on Technological Progress" contains very specific language addressing the questions of new technology, the bargaining unit, and retraining policies:

> It is recognized that advances in technology may alter, modify or otherwise change the job responsibilities of represented employes at plant locations and that a change in the means, method or process of performing a work function including the introduction of computers or other new or advanced technology will not serve to shift the work function from represented to non-represented employes.

> In view of the Corporation's interest in affording maximum opportunity for employes (sic) to progress with advancing technology, the Corporation shall make available short-range, specialized training programs for those employes who have the qualifications to perform the new or changed work, where such programs are reasonable and practicable. Therefore, in the event the work performed by employes covered by the National Agreement is altered as the result of technological changes so that additional short-range training may be required, the Corporation is willing to train such employes where practicable to enable them to perform such work. (Agreement Between General Motors Corporation and the UAW, pp. 431-432)

This statement makes it rather clear that employees will be considered when technical change impacts unfavorably on their job security.

Thirdly, the new agreement also established a Joint Skill Development and Training Committee whose responsibilities (among others) include: (1) seeking ways of arranging for training, retraining and development assistance for employees displaced by new technologies, new productive techniques and shifts in customer product preference; and (2) developing and providing training to enhance skills for present and anticipated job responsibilities and to meet new technology. (Agreement Between General Motors Corporation and the UAW, pp. 277-288)

Furthermore, the charge to this committee is to be backed up by a dual funding commitment by General Motors. There is to be 5 cents contributed for every hour worked to the Executive Board-Joint Activities. This amount will be used to fund all joint efforts, including the Joint Skill Development and Training Committee, the Joint Council for Enhancing

Job Security and the Competitive Edge, and the National Committee to Improve the Quality of Worklife. In addition to these funds, GM has agreed to provide $80 million per year "for current and expanded training for bargaining unit employees." (Agreement Between General Motors Corporation and the UAW, p. 425)

Taking all these elements together, it is reasonable to believe that General Motors will be able to develop the skills it needs for the future primarily from current bargaining unit employees. Assuming that other auto manufacturers follow this lead, it seems sensible to assume that the 3,000 to 5,000 robotics technicians projected for the auto industry in 1990 will represent retrained current employees rather than new hires. The very extensive existing skilled trades apprentice programs in the auto industry also add credibility to this scenario. The General Motors-UAW program appears to serve both the private and public interests in technological change in the auto industry. Technical change and job displacement will be accommodated without compromising the job security of employees unduly.

It remains an open question whether the pattern being established in the auto industry to accommodate the introduction of new technologies such as robotics will be followed in other industries. However, this possibility cannot be dismissed out of hand. Several major unions have recently announced either their intent to bargain over retraining and job security issues or have already signed contracts where these issues played a role. (Ruben, 1982a, pp. 44-45; Ruben, 1982b, p. 44; "A Year of Settling for Less - and Breaking Old Molds," pp. 72-74) To be sure, no agreements to date approach the scope or magnitude of the retraining commitment contained in the auto industry agreements, but these issues are becoming more important to workers.

We also note that it may be a good human resource management to upgrade workers wherever possible as robotics technicians. There is the rather obvious need for a cohesive and cooperative workforce as new technology is introduced. Moreover, as discussed earlier, there will likely be experienced personnel already in the factory who are trained in complex machinery repair and who will not be intimidated by robots. So these experienced workers may be ideal candidates for retraining to accommodate the introduction of robots.

It may be helpful to briefly recap our remarks about the demand for robotics technicians, especially for those readers unfamiliar with the jargon of labor market analysis. We project a cumulative total requirement for 12,300 to 24,600 robotics technicians by 1990, but the auto industry will likely meet its need through what economists generally call the internal labor market, i.e., by retraining current workers to staff the robot maintenance function. From the perspective of the young person seeking training in robotics from a 2-year community college without a prior commitment from an employer (the external labor market), our projections should thus be reduced at least by the anticipated needs of the auto industry. That results in a projection for the external labor market of 9,300 to 19,600 robotics technician jobs. Even these estimates are a maximum, since other industries besides autos will also likely employ a retraining strategy to some extent.

The same interpretation of our projections applies to community colleges offering robotics technician training, except that some community colleges will probably contract with specific employers to provide retraining for their current employees as well as the training they traditionally provide for the external labor market. It is very early in the game to make any judgments about supply adequacy for robotics technicians; the occupation barely exists today. Yet, interest appears extremely high among student populations, and robotics technician curricula are expanding rapidly.

Macomb County Community College in Warren, Michigan is generally acknowledged as the originator of the robotics technician curriculum in the United States. (Schreiber, pp. 78-79) Ms. Schreiber dates the beginnings of Macomb's program to 1978 when they added a specialty in robotics to their fluid power technology associate degree program. Because of all the media emphasis in the last 12 months, Macomb has had to turn away hundreds and hundreds of students who wanted to enroll in robotics courses in the Fall of 1982.

In addition to Macomb, robotics technician curricula are now offered at three other Michigan community colleges (Henry Ford, Oakland, and Washtenaw) and there are planning efforts under way in at least eight more (Grand Rapids, Highland Park, Kellogg, Lansing, Mott, Schoolcraft, St. Clair, and Wayne). Interest is running so high in robotics that when Oakland Community College in Michigan opened a brand new program in the fall of the 1982-83 school year, they immediately enrolled over 600 students in the introductory course. The next semester another 900 were enrolled. Washtenaw Community College, like Macomb, also turned away hundreds of students this year, and interest in Schoolcraft College's planned technician program appears high as well.

Michigan does seem to be the clear leader in the area of robotics technician education presently. Robotics International of the Society of Manufacturing Engineers is currently conducting a nationwide survey of colleges, universities, and corporations involved in education and training for people working with industrial robots. Results of the survey will be published in an education and training directory later this year. Preliminary reports indicate that there is intense interest in robotics technician programs in Tennessee and Texas. There appear to be at least 7 colleges outside the State of Michigan now offering robotics technician curricula. Clearly, the Robotics International survey which will be updated on a yearly basis will offer much needed information to students and educators alike.

There are other robotics technician programs besides those in the community college system. At least two (and probably more) private-for-profit schools are operating now, and we have received inquiries from several others that are considering such a program. These schools appeal especially to the unemployed worker who has some technical background and wishes to become a robotics technician in a short period of time, generally one year or slightly less. In fact, these schools may be appealing to any student where (for whatever the reason) time compression of the training is a key consideration.

There have also been at least two federally assisted robotics technician training programs aimed at the displaced worker. The City of Warren, Michigan sponsored a 40-week program under the Comprehensive Employment and Training Act (CETA) in 1982. Approximately 20 students were enrolled at a cost of 12,000 per student. ("Robotics Class Looks Ahead") The other program was conducted by the Downriver Community Conference, Wyandotte, Michigan, a federal demonstration project dealing with displaced workers. The training was done at Macomb Community College. According to a recent letter announcing the "First Annual Job Fair" for the Downriver graduates, 24 people were enrolled in the program with from 5 to 20 years of prior work experience. The placement results of these pilot programs have been disappointing, but that should not be interpreted as a sign of failure. The graduates appear to be receiving numerous interviews. We are confident that these retrained workers will be hired as soon as the current recession subsides and robot sales resume their healthy growth.

We estimate that two to three hundred students will complete robotics technician curricula in 1983. In the current school year we estimate that there are 2,500 to 3,000 students enrolled in the introductory robotics course at schools that offer a 2-year robotics technician degree. Given the combination of high student interest in robotics and the apparent responsiveness (perhaps overresponsiveness) of the educational system to that interest, enrollment may climb significantly in the next school year, 1983-84. In short, there appears to be no need to worry about a lack of supply of robotics technicians.

Some attention should be given, however, to ensure the quality of supply. A Robotics Clearinghouse project is being sponsored by the Michigan Department of Education to assist in the development of curricula in the automated manufacturing systems/robotics technology area. A consortium of Washtenaw Community College, Henry Ford Community College and Macomb Community College are participating in this effort. They have developed plans for a survey of robot users to help in determining what the needs of potential employers might be. This effort offers the potential to see that the educational product is the right one.

One of the dangers is that students and educators might overconcentrate on robots. This conclusion may seem surprising, but there are several reasons for it. First, robots are only one type of automated equipment, and it is important that these technicians be flexible enough to work on other automated equipment as well. Our interviews revealed rather strong support for broad-based training in the fundamentals of electronics, hydraulics, etc., rather than overspecialization in robots. Second, the demand for robotics technicians will likely be small until the latter 1980s while supply appears to be expanding rapidly now, so some technicians may not find immediate employment in robotics-related fields.

Unfortunately we are not able to adequately assess the prospects of employing robotics technicians in other closely related fields. Clearly, that depends on the type and adequacy of the specific curricula completed. The BLS occupational projections foresee an above average increase in the need for engineering and science technicians in the decade of the 1980s of 24.5 percent to 34.1 percent, and this occupational category is large with over a million members in 1980. But we simply do not know how many of these jobs someone trained as a robotics technician might be qualified to do. For these reasons our advice to students is to avoid overcommitment to a narrowly defined robotics technician curriculum. Likewise, schools should avoid overzealous promises of employment directly in robotics, at least until the market for this emerging occupation becomes more clearly delineated.

Overall supply-demand conditions for robotics technicians are extremely difficult to evaluate now. Robotics itself is just an infant industry, it is unknown how many of today's enrollees will actually complete the curricula, and it is unknown how rapidly additional schools will begin to offer such programs. With those caveats in mind, we attempt to draw some conclusions based on the scattered information available.

In the near term there may be a shortage of technicians. If a vigorous recovery from the recession ensues, demand could pick up overnight. Obviously, supply does not respond as swiftly because of the time required for training. But we must note that there is at least one student enrolled in the introductory robots course in the 1982-83 school year for every robot that will likely be sold in the U.S. during 1983. Moreover, given the high interest in robotics training among student populations and the fact that robotics technician curricula are just beginning to proliferate nationwide, the greatest likelihood in the 1983-84 school year is for new enrollments to grow more rapidly than robot sales. If these trends continue very long, we think most observers would agree that there will likely be a surplus of robotics technicians.

For these reasons we strongly urge providers of education to concentrate on quality rather than quantity. They must ensure that their product is what employers need. The breadth of training is also a very important consideration because of the uncertainties in demand for robotics technicians. We generally prefer the educational approach that adds robotics courses to an electronics technician or other similar training program rather than a more specialized robotics technician program. Students, on the other hand, must understand that the creation of 25,000 robotics technician jobs by 1990 does not mean all of these positions will be available to new labor market entrants. We do not expect to see hiring from outside to staff the robot maintenance function in the auto industry, and it is possible that this will be true for other industries as well.

In the final chapter, the job displacement and job creation projections will be drawn together to describe the very significant skill-twist that appears to be associated with the introduction of robots. Let it suffice at this point to show that most of the jobs created will require a high quality technical education. On the other hand, most of the jobs to be displaced require little formal education. This poor match appears to be a major labor market implication of robots.

5
Summary and Conclusions

Introduction

The robots are coming; not as rapidly as anticipated by some nor with the devastating impact predicted by others, but they *are* coming. Furthermore, we all have a stake in the impending change, at least to the extent that robots will be part of a movement to raise the productivity of American factories and retain the competitiveness of American goods in national and international markets. We have argued throughout this monograph that robots should be regarded simply as another labor-saving technology, one more step in a process that has been going on for some 200 years.

This study has focused on the human resource implications of the introduction of industrial robots, but to begin it was necessary to put the so-called "robotics revolution" into some perspective. Hard data about industrial robots are scarce today. Most of the public awareness of robots has been shaped by the hyperbole in the popular press. Futurists and others compete for media attention with wild projections of the impacts of robotics—800,000 people making robots, 1.5 million technicians maintaining robots, and millions of workers displaced—with little or no consideration of the practical issues involved. We believe the intense media attention on robotics in the past year or so may have seriously confused the issues.

First, we submit that the very use of the word "revolution" is inappropriate when dealing with any manufacturing process technology. Capital goods for production have long lives and are not scrapped immediately when something better comes along. Numerically controlled machine tools, usually regarded as the capital equipment most closely related to robots, expanded at a growth rate of only 12 percent for the most recent 10-year period. After 25 years, only 3 to 4 percent of all metalcutting machine tools are numerically controlled. Even digital computers, widely heralded as the most significant technological innovation of the 1960s and 1970s, expanded at a growth rate of only 25 percent. Yet many are implicitly assuming much higher growth rates for industrial robots. In terms of actual application, all process technology changes are evolutionary rather than revolutionary because of the physical, financial and human constraints on the rate of change of process technology.

Second, the fear of massive unemployment caused by the introduction of industrial machinery appears to be unfounded. Such fears began with the dawn of the industrial era in the 1700s. They are particularly acute during major recessions. For example, the "automation" problem was of urgent national concern in the early 1960s after a halting recovery from the sharp recession of 1958-59. There were grim predictions that automation was causing permanent unemployment in the auto industry and other industries. A national commission was appointed to study the problem and in 1966, with the economy near full employment, the commission rendered its final report. To no one's surprise, they concluded that a sluggish economy was the major cause of unemployment rather than automation.

Third, there appears to be a fundamental lack of understanding that the association of technological change, economic growth, and job displacement is not just a coincidence; they are intertwined and inseparable. That is not to imply that adoption of new technologies necessarily insures economic growth, or that displaced workers will always find new jobs. However, it does mean that we all have a vital stake in productivity gains (i.e., in displacing jobs) because that is what allows the *possibility* of economic growth. The price of a growing, dynamic economy that makes more goods and services available to all of us is job displacement, or the elimination of jobs through technological change.

The intent of this study has been to provide an informed, balanced review of the direct impact of robots on the employment picture in the U.S. and Michigan between now and 1990. Given the lack of universally accepted data about robots, and a robot industry that is still in the formative stage, it was necessary to resort to considerable projection and estimation. This creates the opportunity to be extravagant; we have tried to avoid this. We have selected the conservative, but realistic alternative wherever there was a choice. By laying all assumptions before the reader, we hope to make that point clear.

This method also has the advantage of focusing disagreement on the particular assumptions used in the study, thus providing the opportunity for refinements or improvements. Our hope is that this study will help restore reason and balance to the discussion of these issues.

Findings

The projections of occupational impact in this study are the result of first forecasting the U.S. robot population by industry and application areas. This approach constrains the employment impacts to reflect the actual expected sales of robots. In this way a consistent economic framework is established within which it is possible to estimate not only the population of robots and job displacement but also the job creation resulting therefrom. This consistency is also very helpful in avoiding unrealistic or exaggerated conclusions.

We expect strong growth in the utilization of industrial robots in the decade of the 1980s. In chapter 2 we forecast that the total robot population in the U.S. by 1990 will range from a minimum of 50,000 to a maximum of 100,000 units. Given our estimate of the year-end 1982 population of approximately 6,800 to 7,000 units, that implies an average annual growth rate of between 30 and 40 percent for the eight years of the forecast period, or roughly a seven to fourteenfold increase in the total population of robots.

This range is intended to contain the actual robot population with a high probability level, and allows for variation in interest rates, capital investment climate, auto industry recovery, and rate of economic growth. We are confident this range will contain the 1990 robot population. That means we do not expect developments such as the total collapse of the automobile industry, a major renaissance in the U.S. capital investment, the early development of a significant number of nonmanufacturing robot applications, or the widespread adoption of robotics technology by small firms.

The U.S. population of robots is developed separately for the auto industry and all other manufacturing. This is partly to take advantage of the fact that the auto producers have announced goals for robot installations which could be factored into our robot population forecast. It also reflects the fact that the major impact of robots in the State of Michigan will be in the auto industry. Our forecast sees 15,000 to 25,000 robots employed in the U.S. auto industry by 1990.

Utilizing the robot forecast by industry, and the assumption of a gross displacement rate of two jobs per robot which was strongly supported in our interviews, estimates of gross job displacement can be derived. We estimate that robots in the U.S. will eliminate between 100,000 and 200,000 jobs by 1990. From 30,000 to 50,000 of these will be in the auto industry, while 70,000 to 150,000 jobs in other manufacturing industries will also be eliminated.

In addition to the assignment of robots by industry, it was necessary to forecast the applications for which they will be used. This is required if the robot population forecast is to be useful in predicting occupational displacement. Otherwise there is no way to connect the robots with the work content of specific jobs. The application areas used in this study are welding, assembly, painting, machine loading and unloading, and other.

When the robot forecast by application area and industry is matched against an occupational data base similarly organized, specific occupational displacement rates can be estimated. In chapter 3 it was shown that while the maximum overall job displacement rate in manufacturing of 1 percent through 1990 is not particularly problematical, specific industry and occupation displacement rates are very significant, even dramatic.

To begin with, the displacement rate derived for the auto industry ranged from 4 to 6 percent of all employment. But when displacement was calculated only against the production workers in the auto industry, the magnitude of displacement was from 6 to 11 percent. Even when considered to be over a period of a decade, these rates of job displacement *are* significant.

When specific occupational displacement rates are calculated, even more striking results emerge. Our results suggest that between 15 and 20 percent of the welders in the auto industry will be displaced by robots by 1990. Even more dramatically, between 27 and 37 percent of the production painter jobs in the auto industry will be eliminated by 1990. While displacement results are generally less significant for specific occupations in all other manufacturing, it is projected that 7 to 12 percent of the production painter jobs there will be lost in the same time frame.

The conclusion of the job displacement estimates is that while job displacement due to robots will not be a general problem before 1990, there will clearly be particular areas that will be significantly affected. Chief among these will be the painting and welding jobs for which today's robots are so well adapted. Lesser impacts will be apparent on metalworking machine operatives and assemblers. Geographically, states such as Michigan, especially the southeastern quadrant with its heavy dependence on autos, will suffer greater displacement than other states or regions.

We do not believe that this job displacement will lead to widespread job loss among the currently employed, however. Even in the auto industry, voluntary turnover rates historically have been sufficient to handle the reduction in force that might be required. In addition, the new General Motors-United Auto Workers contract seems to provide adequate job security assurances, and the retraining commitment necessary to back them up. Thus we do not expect any substantial number of auto workers to be thrown out of work due to the application of robots. Any unemployment impact is likely to be felt by the unskilled labor market entrants who will find more and more factory gates closed to the new employee. Therefore, if there is an increase in unemployment as a result of the spread of robotics technology, we fear the burden will fall on the less experienced, less well educated part of our labor force.

Turning our attention to the job creation issue, in chapter 4 we forecast the direct creation of about 32,000 to 64,000 jobs in the U.S. by 1990 in four broad areas: robot manufacturing, direct suppliers to robot manufacturers, robot systems engineering, and corporate robot users. The jobs in corporate robot users identify maintenance requirements for robots, while the jobs in robot systems engineering identify the applications engineering requirements for robot systems, without regard to industry of employment.

In these projections we assumed that the status quo would be maintained in both the import and export markets for robots, primarily because of a lack of any better information. But there is certainly no guarantee that U.S. producers will maintain their share of the national or worldwide market. This threat is especially menacing because of Japanese and European expertise in robotics technology.

The projections of robot-related job creation by occupation are very speculative because of the limited experience to date with robots and the uncertainties involved in predicting the future occupational profiles of firms that do not yet exist. However, the high technical component of labor demand is quite startling. Well over half of the jobs created will require two or more years of college training.

The largest single occupational group of jobs created by robotics will be robotics technicians. This is a term which is just coming into general usage; it refers to an individual with the training or experience to test, program, install, troubleshoot, or maintain industrial robots. We anticipate that most of these individuals will be trained in community college programs of two years duration. We project that jobs for about 12,000 to 25,000 robotics technicians will be created in the U.S. by 1990. We do not anticipate a supply problem for robotics technicians, as the community college system gives every indication that they will be ready and willing to train whatever numbers are needed. In fact, our current concern is that they may, in some instances, be increasing the supply too rapidly.

In the auto industry, we expect the robot maintenance requirement will continue to be met by the members of the UAW Skilled Trades Council. General Motors already has agreed to a retraining effort approximating $120 million annually. We believe the strong implication of the contractual arrangements is that auto industry employers will not be required to hire from the outside to meet their robotics technician needs. Other major robot users may follow the lead of the auto industry, but it is impossible to predict that with assurance at this early date.

There also will be a relatively large number of graduate engineers needed to implement the expansion of robotics technology in U.S. industry. We estimated the requirement from about 4,600 to 9,300 new engineers. While these numbers are comparatively small, only one-fifth of one year's production of engineers at the baccalaureate level, there is already a clear shortage of engineers, so we start from a deficit position. In addition, we face the challenge of other likely engineering demand increases as well as the historical instability of engineering enrollments. Thus it is quite likely that a shortage of engineers could compromise the expansion of robotics technology.

The most remarkable thing about the job displacement and job creation impacts of industrial robots is not the fact that more jobs are eliminated than created; this follows from the fact that robots are labor-saving technology designed to raise productivity and lower costs of production. Rather, it is the skill-twist that emerges so clearly when the jobs eliminated are compared to the jobs created. The jobs eliminated are semi-skilled or unskilled, while the jobs created require significant technical background. We submit that this is the true meaning of the so-called robotics revolution.

Implications

This study has focused on the employment and training implications of the spread of robotics technology by the year 1990. It is probably fair to say that the major determinant of the overall impact of robotics in the '80s is the fact that robotics is an infant industry today. There is no way that the robotics industry can grow to be a giant in less than a decade (the futurists notwithstanding). It has repeatedly been demonstrated in this study that even with an extremely rapid growth rate of 30 to 40 percent annually in the population of robots in American industry, the robotics industry will still be small in 1990. The consensus prediction of the size of the industry in 1990 is $2 billion of sales annually. But General Motors Corporation alone had sales of nearly $50 billion in the U.S. in 1981. Chrysler had net sales of nearly $11 billion in a depressed economy. So an industry with $2 billion sales will still be very small in 1990.

The growth of the industrial robot population will not be restricted because of the inability of manufacturers to produce robots fast enough; there is plenty of capacity today and we are confident it can be expanded rapidly. The limits on the use of industrial robots will derive from the human, financial, and physical constraints that retard changes in manufacturing process technology. We have argued that process technology is significantly different from product technology. Robots cannot spread through America's factories the way Rubik's cube spread through America's homes. We have demonstrated by analogy with other process technology innovations that such change is evolutionary rather than revolutionary. To repeat a phrase used earlier, we believe the very use of the word "revolution" is inappropriate when dealing with *any* manufacturing process technology. Nevertheless, this examination of the human resource implications of the rapid growth in the robot population up to 1990 has revealed some potentially significant problems.

First, while we are convinced that there will be no general worker displacement problem, there clearly *will* be particular pockets of displacement that may cause labor market distress. Particular occupations, industries, and locations will suffer the brunt of the job displacement impact. Examples include industrial welders and production painters, the auto industry, and Southeast Michigan. In each of these cases, substantial job displacement will occur in the decade of the '80s because of the application of robots. While a review of labor force attrition rates suggests that there will be very few workers actually thrown out of work even in these highly impacted areas, there is still some potential for displaced workers in these situations. We do not pretend our results are precise enough to make such calls with unfailing accuracy.

Robotics is obviously not the only change that will be forthcoming in the rest of the decade. There will be many influences on the levels of employment by occupation and industry. We have only examined the impact of robots, ignoring any other effects. This includes possible expansion in volume of production due to price reductions or quality improvements. We also ignored potential international trade implications of robotics technology. In essence we have imposed our assumptions about the robot population and job displacement on an existing economic structure, without allowing for the natural adaptation and feedback effects that will likely occur.

In addition, we are very sensitive to the fact that we do not begin from a satisfactory employment situation. We still languish in the trough of a severe recession; aggregate unemployment rates are setting post-Depression records. Discussion of even minimal job elimination in the next few years as a result of the application of industrial robots seems particularly grim in times like these. We need more job creation, not job elimination. Even though it is ludicrous to believe that the seven thousand robots now operating in American factories have played a significant causative role in the unemployment of 11 million Americans, job loss hysteria has reached a point where there is a need to find scapegoats for our desperate situation.

Auto workers particularly are caught in a difficult trap. If it is true that the greater incidence of robots (and the manufacturing quality they help provide) plays a role in the success of Japanese automobiles in the American market, the challenge of robotics must be met. But the introduction of robots will clearly cause the direct displacement of some auto worker jobs. It is impossible to guarantee that robots will help regain some of the market share lost to the Japanese and therefore result in the restoration of jobs previously eliminated through competitive pressure. We do not know whether the Japanese challenge will be met successfully. Nor do we know how important robots may be in meeting this challenge. We do believe that the robots are coming to the auto industry anyway and must be accommodated. Those opposing technological progress rarely change the course of history for long.

Nevertheless, we believe it is clear that the rapid spread of robotics technology through American industry in this decade will not throw any significant number of American workers out of their jobs. Therefore we do not feel compelled to call for a major policy response to a problem that does not exist. Robots may add somewhat to our existing displaced worker problems during the 1980s, but they will not be a major contributor. Whatever policy initiatives are designed for the general displaced worker problem should adequately address those displaced by robots as well. We do believe that targeting such efforts occupationally, industrially, and geographically should be an important consideration in the design of any program to address the displaced worker problem.

The second major conclusion of the study is the skill-twist that characterizes the jobs displaced and the jobs created by robotics. Even though it is difficult to predict the exact occupational structure of an infant industry, we think it is clear that robotics will employ workers who are significantly more skilled on the average than more traditional industries. As shown earlier, over half of all the jobs created by robotics will require a 2-year degree or more. The new jobs will require much more technical background than manufacturing jobs in the past. The major implication of this observation is that retraining the workers displaced by robots for the new jobs created may not be realistic. On the other hand, our results suggest that the pace of displacement will be sufficiently gradual that human resource planning can obviate the problems.

To use the auto industry as an example again, it would be difficult to retrain a welder from the line to repair and maintain the welding robot that will be doing his job in the future. However, it is not particularly difficult to train skilled plant maintenance workers to also maintain industrial robots. Thus the most efficient human resource management strategy may involve retraining the former welder to operate a machine which will not be robotized, while the robotics training is concentrated on those workers who are skilled already. it is not likely that the very same person replaced by the robot will be doing the new job or jobs created by the robots. Of course, the net result of such retraining and upgrading will be a markedly different skill mix; in other words, the skill-twist.

We have also expressed our concern about the job outlook for unskilled youth in the future. We believe it is likely that employment in manufacturing will continue to expand much more slowly than the labor force as a whole. To the extent that we already have a serious job deficit for unskilled youth, the growth of robotics will tend to exacerbate the problem. As displaced unskilled and semi-skilled workers are retrained and transferred to the remaining blue collar jobs in the factories, the outlook for hiring new unskilled workers declines correspondingly. Thus, we fear that any unemployment burden caused by robotics will ultimately fall on the younger generation.

It may be a fortuitous coincidence of the baby ''bust'' that the number of youth entering the labor market will be declining substantially at approximately the same time as the job displacing impacts of robotics become significant. Nevertheless, we urge young people to get a solid science and mathematics background if they want to be employable in the manufacturing sector.

The third major thrust of the study is the question of supply of the technical skills required by robotics technology. We have identified two very different problems, a *potential oversupply* of robotics technicians and a *probable shortage* of engineers. The rapid spread of robotics technology will enhance the demand for engineering talent, adding to an existing shortage situation. While robotics alone will not impact significantly on the demand for engineers, we believe there are other reasons for expecting the shortage to grow more serious during the rest of the decade. Thus we add our voices to those calling for immediate national attention to the supply of engineers.

The supply problem of robotics technicians may well turn out to be that of oversupply. We believe that the growth of the robotics industry will be very rapid, but it seems clear that student interest and the ability of the community college system to increase the supply is growing even more rapidly.

A continuation of the expansion of the last year or so in course offerings and enrollments on a national scale will very quickly swamp the ability of the industry to absorb trained people. For that reason, we endorse careful attention to the breadth of training. A firm grounding in theory and general principles of electronics, controls, hydraulics, etc. will stand the graduates of such programs in good stead whether they actually work primarily with robots or not.

In addition, we became convinced during the course of the study that there is an unmet need for manufacturing technology generalists, both at the graduate engineer and technician levels. A number of experts familiar both with manufacturing technology and the capabilities of todays engineering graduates complained about the over-specialization of training provided. Rebuilding American industrial strength will require individuals trained to be familiar with manufacturing technology in a broad sense. It is asserted that both Japan and Germany have programs for training such people. This may be one of the keys to increases in manufacturing productivity in these economies.

The last major issue to be addressed is "What comes after 1990?" Clearly the implication of our assumption of exponential growth in the robot population is that the job displacement effects are growing exponentially as well. The examination of job displacement in the single year 1990 in chapter 3 illustrated this effectively. If robots could eliminate one job opening of eight projected for production workers in manufacturing in 1990, when does it reach one in two, or one in one?

While it is a simple matter to extend the calculations and generate an answer to this question, we think it is an exercise that should be done with extreme caution. Using the assumptions of this study, we could forecast a robot population of 250,000 to 500,000 for the U.S. by the end of 1995. We also think it would be irresponsible to do so at this early date. The data base does not exist and we do not believe the potential policy responses to such implied levels of displacement require lead times of more than three or four years. Thus the wisest course would be to monitor robotics developments for the next few years, keeping our vision fixed on a target six to eight years ahead. This strategy provides sufficient decision-making time while simultaneously maximizing the quality of information available at the decision point.

Nonetheless, we think it is possible to anticipate some general trends that lie ahead for the manufacturing workforce. The evolution of manufacturing process technology will undoubtedly continue. Productivity enhancing investment in robots and other new technology will go ahead. Rising productivity in manufacturing will cause a continued decline in the proportion of American workers employed in the manufacturing sector, even if the challenge of foreign imports is met. We believe that the skill-twist demonstrated in this study can probably be generalized to other manufacturing technology developments. Thus we believe it is possible to predict a continued decline in manual, semi-skilled jobs while the new jobs created will be increasingly technical and scientific.

It should also be reiterated that some of the substitution of machines for human labor can and will be regarded as a blessing. There are a great many dirty and dangerous jobs that robots or other machines could do effectively, thereby preventing human exposure to these situations. Provided meaningful alternative work can be found for the occupants of those jobs, there is no need to feel remorse at the loss. We should not be so blinded by our short term economic problems that we forget the connection between productivity, job displacement, and economic progress.

Finally, there is no reason to believe that the addition of robots to our factories is anything other than an evolutionary change. Industrial robots are simply one more piece of automated industrial equipment, part of the long history of automation of production. Robots will displace workers in the same way that technological change has always displaced workers. There is a possibility that this job displacement will be a significant problem, particularly in a given occupation or industry or geographical area. There is also the certainty that robots will create new jobs. Most of these will be quite different from the kinds of jobs eliminated. Robotics may challenge our ability to manage our most valuable resource, but there is no reason for the job displacement or the skill-twist impacts to create tragic consequences. It is not time to panic; it *is* time to begin rational planning for the human resource implications of robotics.

Annotated Bibliography

Agreement Between General Motors Corporation and the UAW. March 21, 1982 (Effective April 12, 1982).
The legal contractual agreement between the UAW and GM. It includes the provisions of the recent negotiations concluded on March 21, 1982 and all other active contract provisions.

Albus, James S. "Industrial Robot Technology and Productivity Improvement," in *Exploratory Workshop on the Social Impact of Robotics: Summary and Issues.* Office of Technology Assessment, U.S. Government Printing Office, Washington, DC, February 1982, pp. 62-89.
Provides an analysis of the technical problem areas retarding the growth and the utilization of robots, summarizes on-going university, non-profit, private industry and government research in robotics, and discusses likely future implementation of robots. The primary reason the implementation of the technology has been slow in the U.S. is the fear that unemployment would become a serious problem. Albus says that fear is misplaced. Work is not fixed in amount and we can always create new work.

Alexander, Tom. "Practical Uses for a Useless Science," *Fortune,* May 31, 1982, pp. 138-145.
Presents a nontechnical discussion of the on-going research and development of artificial intelligence (AI). Notes that AI is important to the development of "smart robots" and automated manufacturing systems. For the most part, AI still requires computer logic that is far beyond the capabilities of current systems, although AI systems are currently available for carefully defined and limited tasks.

Anderson, Arthur and Company, et al. *U.S. Automotive Industry Trends for the 1980s.* Chicago, 1979.
Using the Delphi forecasting method, this report surveyed industry experts to develop a consensus as to future technological, market and management trends for the 1980s. The findings of the survey cover six areas including technological trends, government regulations, the

market outlook for automobiles and parts, the investment climate, the labor market outlook and the development of a Strategic Planning Strategy for the 1980s.

———. *U.S. Automotive Industry in the 1980s: A Domestic and Worldwide Perspective.* Chicago, July 1981.
This study utilizes the Delphi method of forecasting and expands upon the 1979 forecast provided by this company to predict trends in markets, technology, productivity improvement, capital investment and government policies for the 1980s and early 1990s.

Aron, Paul. "Robots Revisited: One Year Later," in *Exploratory Workshop on the Social Impacts of Robotics: Summary and Issues.* Office of Technology Assessment, U.S. Government Printing Office, Washington, DC, July 1981, pp. 27-61.
Aron's second evaluation of robotics (both reports, No. 22 and 25, are available from Daiwa Securities America Inc.) concludes that Japanese robotics growth tends to be understated. Aron predicts that by 1990 the U.S. will have an annual market of 21,575 robots, while the Japanese market will be 57,450.

———. "Robotics in Japan: Past, Present, Future." Presentation to Robots VI Conference, March 1982.
A discussion of the successful production and utilization of industrial robots in Japan. Aron states that this can be attributed to the employees' acceptance and active encouragement of robotics application. Japanese employers are responsible for retraining displaced workers, and Japanese employees are provided a share of the profits. In many cases robot users in Japan manufacture their own robots.

Ascher, William. *Forecasting: An Appraisal for Policy-Makers and Planners.* Johns Hopkins University Press, Baltimore, 1978.
Surveys and evaluates the "state of the art" in forecasting. It includes the Delphi method as a part of technological forecasting.

"A Year of Settling for Less - and Breaking Old Molds," *Business Week,* December 20, 1982, pp. 72-74.
Describes 1983 as a year that will continue the trend toward modest wage increases in collective bargaining agreements. Nonmonetary factors are being emphasized more often, such as job retraining and job security.

Ayres, Robert and Steven Miller. *Understanding and Managing the Productivity, Human Resource and Societal Implications of Robotics*

and Advanced Information Systems. The Robotics Institute, Carnegie-Mellon University, April 1982a.
The author asserts that the application of robotics technology needs to be developed in such a way as to insure economic competitiveness and growth while providing employment opportunities for the labor force. In order to manage a successful transition the authors have identified a broad set of analytical concerns, potential impacts and policy issues to be addressed over the years.

———. "Industrial Robots on the Line," *Technology Review,* May/June 1982b, pp. 35-46.
This published paper is a shortened version of the unpublished paper, *Impacts of Industrial Robots.* (The Robotics Institute, Carnegie-Mellon University, November 1981.) The authors conclude that neither government nor private industry is adequately preparing workers whose jobs are most likely to be eliminated by robots. The authors suggest a number of avenues for retraining displaced workers.

———. "Robotics, CAM, and Industrial Productivity," *National Productivity Review,* Vol. 1, No. 1, Winter 1981-82, pp. 42-66.
The authors conclude that the primary direct impact of robotics and CAM will be to significantly boost the output capacity of batch production operations. Although it is unclear how long it will take to achieve the "unmanned factory of the future," demand will remain very high for at least a decade or more for robots and other computer-controlled machines.

———. *The Impacts of Robotics on the Workforce and Workplace.* Carnegie-Mellon University, Department of Engineering and Public Policy, June 1981a.
Reports the results of the original student project in which Ayres and Miller were the principal investigators. Estimates of potential job displacement caused by robots were based on a survey of corporate users. Concludes that today's robots could theoretically replace 1 million operatives in manufacturing and that the next generation of robots could potentially replace an additional 3 million operatives in manufacturing.

———. *Preparing for the Growth of Industrial Robots.* Carnegie-Mellon University, Department of Engineering and Public Policy, September 1981b.
Given the potential for robotization in metalworking occupations, this paper outlines the intervention necessary to prepare the affected

workers and thereby prevent social trauma. In so doing the authors describe what they feel to be the roles of private industry, educational institutions, unions and government.

Ayres, Robert, Leonard Lynn and Steven Miller. *Technology Transfer in Robotics: U.S./Japan.* Carnegie-Mellon University, Department of Engineering and Public Policy, October 1981.
Discusses differences between U.S. and Japan in development and introduction of robotics technology. Japan provides more governmental support, Japanese industries tend to make their own robots in-house, and workers are provided lifetime security in major industries.

Backman, Jules, ed. *Labor, Technology, and Productivity in the Seventies.* New York University Press, New York, 1974.
Issues discussed in this collection of papers include the trade union movement in the U.S., its relationships with business, and the momentum and trends of technological progress.

Battelle-Columbus Laboratories. *The Development of High Technology Industries in New York State,* six special reports. New York State Science and Technology Foundation, Albany, 1982.
This study assesses the environment in New York State for the attraction, expansion and retention of high technology industries. The analysis takes into account the strengths and weaknesses of the state, local manpower, and financial incentives as well as the major actors and programs necessary to implement an action-oriented strategy by all decisionmaking sectors in the state.

———. *Development of the Robotics and Biotechnology Industries in the State of Michigan,* Mimeographed. Michigan State Department of Commerce, November 1981.
This study was conducted to determine the factors that attract the robotics and biotechnology industries to locate in particular geographical areas. The results of interviews conducted with robotics and biotechnological firms indicate that Michigan has a well developed industrial base for both of these high technology industries. The state needs to set policies for centralizing R&D resources in addition to policies directed at improving the business climate in Michigan.

Behuniak, John A. *Economic Analysis of Robot Applications.* Society of Manufacturing Engineers, Technical Paper MS 79-777, 1979.
Studies the economics of robot application analyzing the payback, return on investment, discounted rate of return and cash flow of three robot applications.

———. "Planning the Successful Robot Installation," in *Robotics Today,* Summer 1981, reprinted in *Robotics Today '82 Annual Edition,* Society of Manufacturing Engineers, Dearborn, MI, 1982, pp. 23-24.
Stresses that careful planning is necessary to successfully integrate robots into the workplace. This includes human considerations as well as economic costs.

Birch, David L. "Presentation to Michigan, Minnesota, Wisconsin Symposium on Small Business." Green Bay, WI, March 1982.
Discusses the growth of the service sector as opposed to the manufacturing sector. Suggests that our economy is becoming one which is largely knowledge based. Concludes that Michigan is an attractive site for high-technology industries.

———. *The Job Generation Process.* MIT Program on Neighborhood and Regional Change, Cambridge, MA, 1979.
Using Dun & Bradstreet data, Birch concludes that all regions tend to experience the same rate of job loss—about 8 percent per year—but regions differ in the number of new firms, births and expansions. Furthermore, the bulk of net new jobs—about 60 percent—are created by small, independent firms with less than 20 employees.

Bowen, Howard R. and Garth L. Mangum, eds. *Automation and Economic Progress.* Prentice-Hall, Inc., Englewood Cliffs, NJ, 1966.
Excellent overview of the work of the National Commission on Technology, Automation and Economic Progress. It includes an abridged version of the Commission report plus selections from 10 of the 41 supplementary studies.

Brown, Lynn E. "A Quality Labor Supply," *New England Economic Review,* July/August 1981, pp. 19-36.
Suggests that the quality of the New England workforce may have been significant in that region's resurgence in the 1970s, especially in the area of high-technology industry.

Buckingham, Walter. *Automation: Its Impact on Business and People.* Harper and Brothers Publishers, New York, 1961.
Assesses the impact of automation (labor-saving technology) on firms and employment. Presents a somewhat pessimistic analysis of the future.

Burford, Roger L. and Joseph L. Katz. "A Method for Estimation of Input-Output Type Output Multipliers When No I-0 Exists," *Journal*

of Regional Science, Vol. 21, No. 2, 1981, pp. 151-161.
Establishes a statistical technique to estimate the effect of a given industry on other industries in a region (the interindustry effect) when only a minimum of information is available. Information is required for the average material purchases of all firms in the region and for the particular industry of interest to utilize the technique.

"Business Outlays To Rise Modestly in '83 First Half," *Wall Street Journal,* December 10, 1982.
Survey results indicate slight increase in capital spending for businesses for first half of 1983. However, when adjusted for inflation, capital spending is expected to continue to decline.

"Capital Spending's Sickening Fall," *Business Week,* November 15, 1982, pp. 36-37.
Reports results of the 29th annual survey of plant and equipment conducted by McGraw-Hill publications indicating that capital spending is expected to drop at a real rate of 8.5 percent for 1983.

Carey, Max L. "Evaluating the 1975 Projections of Occupational Employment," *Monthly Labor Review,* June 1980, pp. 10-21.
This article compares BLS projected 1975 occupational employment with actual employment estimated from CPS data. Error rates for specific occupations were quite large. Hope was expressed that the new OES data base would improve the quality of the occupational projections.

———. "Occupational Employment Growth Through 1990," *Monthly Labor Review,* August 1981, pp. 42-55.
Summarizes for the first time occupational growth to 1990 using the OES data base rather than census based data.

Carey, Max L. and Kevin Kasunic. "Evaluating the 1980 Projections of Occupational Employment," *Monthly Labor Review,* July 1982, pp. 20-30.
This article compares the 1980 occupational employment projections (made in 1970) with 1980 actual levels (estimated from CPS data). The average error for 64 detailed occupations was 22 percent in absolute levels of employment. Includes an interesting discussion of the difficulties of making such projections.

Cetron, Marvin and Thomas O'Toole. "Careers With a Future: Where the Jobs Will Be in the 1990s," *The Futurist,* June 1982a, pp. 11-19.
Predicts that the world's job markets will change dramatically in the

next 20 years. The major new jobs in the 1990s are predicted to be in robot technology, computer technology, laser technology, energy technology, waste technology, and housing rehabilitation.

———. *Encounters with the Future: A Forecast of Life into the 21st Century.* McGraw-Hill Book Company, New York, 1982b.
This wide ranging forecast describes trends in economics, politics, social affairs, health, science and technology for the next twenty years. The forecasts provide advice on all aspects of life for the 21st century, including the jobs of the future.

Choate, Pat. *Retooling the American Work Force.* Northeast-Midwest Institute, Washington, DC, July 1982.
This largely descriptive monograph concludes that a national strategy for retraining is mandatory in the 1980s. Millions of workers will be displaced by technological change and other factors, while the transfer to new jobs requires new skills.

Committee for Economic Development. *Stimulating Technological Progress.* Heffernan Press, Inc., Washington, DC, January 1980.
Concludes that the reason for lagging U.S. investment and technological progress are public policies which discourage new capital investment. The committee therefore recommends policy changes such as taxes which increase the return on investment, reformation of patent policies, reducing regulatory controls and federal support of R&D.

Conigliaro, Laura. *Robotics newsletter,* Numbers 1-11. Prudential-Bache, Inc., New York, 1980-82.
A continuing newsletter which provides financial and investment advisory information on the robotics industry.

Connole, Anthony W. *Industrial Robots: A View of Future Social Implications.* Eikonix Corporation, Burlington, MA, December 1976.
Attempts to dispel the myth that unions oppose automation and technological change. Suggests that the introduction of robots will take place when the health of the general economy can best absorb the labor displacement.

Dizard, John W. "Giant Footsteps at Unimation's Back,' *Fortune,* May 17, 1982, pp. 94-99.
Hints that Unimation may have difficulty maintaining its dominant position in the robotics industry in the face of entry by such firms as IBM, Bendix, GE and United Technologies.

Doan, Herbert D. "The Development of Technology-Based Businesses in Michigan." Presented to the 3rd Annual Growth Capital Symposium. Ann Arbor, MI, March 1982.
Describes the activities in the state aimed at the development of technology-based business, including the funding foundations, research and training institutions and corporations.

Donovan, Raymond. "Remarks of the Secretary of Labor on Improving Productivity Growth." Presented to the Productivity Advisory Committee, Washington, DC, January 1982.
Discusses the factors contributing to the decline of productivity growth in the U.S. and the need for improving both the efficiency and growth of productivity. Indicates that the advent of high technology will create changes in both the supply and demand for labor, especially in skill requirements.

Drucker, Peter F. "The Next American Work Force: Demographics and U.S. Economic Policy," *Commentary,* October 1981, pp. 3-10.
The U.S. will experience an irreversible decline in the manpower necessary to maintain its blue-collar labor force. The new labor force entrants will be fewer (one-third of the supply available in the last two decades) and better educated. The challenge is to capitalize on a highly educated labor force.

Engleberger, Joseph L. *Robotics in Practice.* American Management Association, AMACOM Press, New York, 1980.
This book is described as the first handbook for managers and engineers on the use of robots in industry. Case studies are provided detailing the use of robotics in several applications such as die casting, welding and forging. Provides assistance to the manager and engineer in judging whether robots are suitable for specific jobs.

Etzioni, Amitai. *An Immodest Agenda: Rebuilding America Before the Twenty-First Century.* McGraw-Hill Book Company, New York, 1983.
Analyzes the next 20 years in terms of social, cultural and economic policy in the U.S. Reaganomics and its likely impact on the infrastructure, capital formation, R&D, energy and human capital are discussed among other topics.

Fechter, Alan. *Forecasting the Impact of Technological Change on Manpower Utilization and Displacement: An Analytic Summary.* Contract Report No. 1215-1, The Urban Institute, Washington, DC, March 1974.

Discusses the state-of-the-art in determining future employment impacts of new technology. Finds very serious problems in forecasting technological change, in isolating technological change from other influences on the labor market, and in the lack of an adequate method for disaggregating labor input.

Fishkind, Henry and R. Blaine Roberts. "Two Methods of Projecting Occupational Employment," *Monthly Labor Review,* May 1978, pp. 57-58.
Compares the OES occupational employment projection method with a fully structured econometric model of the State of Florida. Conclusions are that the econometric model performs slightly better, but not enough to justify the additional cost and effort in maintaining the model.

Fisk, John D. *Industrial Robots in the United States: Issues and Perspectives.* Report No. 81-78 E, United States Congressional Research Service, March 1981.
Discusses industrial robots, their present and future use in American industry and the range of possible effects this technology may have on industry and the workforce. Examines "adjustment mechanisms" for the potentially displaced workforce.

Flaim, Paul O. and Howard N. Fullerton, Jr. "Labor Force Projections to 1990: Three Possible Paths," *Monthly Labor Review,* December 1978, pp. 25-35.
Summarizes the BLS projections of the labor force in 1985 and 1990. The three different growth scenarios differ primarily in their assumptions about the rates of change in participation rates for women and black men. All projections foresee a reduction in youth in both absolute and relative terms.

Freeman, Richard B. *The Over-Educated American.* Academic Press, Inc., New York, 1976.
Analyzes the market for college graduates in the mid-1970s, the first time in many years that new college graduates were having difficulty in obtaining college-level jobs. Among other things, Freeman documents the historical volatility in engineering enrollments and suggests that it is likely to continue.

Gerstenfeld, Arthur and Robert Brainard, eds. *Technological Innovation: Government/Industry Cooperation.* John Wiley and Sons, New York, 1979.
A collection of papers by authors from nine countries which were

presented to an international conference. The basic research question was: Can industry and government cooperate to guide and stimulate technological innovation?

Gevarter, William B. *An Overview of Artificial Intelligence and Robotics,* Volume II, *Robotics.* U.S. Department of Commerce, National Bureau of Standards, Washington, DC, March 1982.
Describes and classifies different industrial robots and their functions. Discusses the present as well as the likely future direction of robotics.

Gold, Bela. *Productivity, Technology, and Capital.* D.C. Heath and Company, Lexington, MA, 1979.
Presents analyses of productivity and technological change at the firm and industry level based upon the author's more than 25 years of research experience. Key conclusions include (1) the need for a more comprehensive framework to analyze productivity changes, (2) the heterogeneous nature of industries cannot be ignored, and (3) actual productivity gains tend to be at wide variance with expectations.

———. *Improving Managerial Evaluations of Computer-Aided Manufacturing.* National Academy Press, Washington, DC, 1981a.
Part of a project to develop a model for managerial evaluation of CAD/CAM systems that will be effective in assessing the distinct capabilities and requirements of these systems. The model is designed to more efficiently estimate the benefits of CAD/CAM as opposed to the standardized budgeting procedures typically used for evaluating the acquisition of new equipment.

———. "Robotics, Programmable Automation and Increasing Competitiveness," in *Exploratory Workshop on the Social Impacts of Robotics: Summary and Issues.* Congress of the United States, Office of Technology Assessment, U.S. Government Printing Office, Washington, DC, 1981b, pp. 91-117.
Concludes that the actual economic impact of major technological changes have usually been less than expected due to an overconcentration on the change itself which neglects the total production framework and its many interactions. Robotics should be evaluated as part of a system of programmable automation for manufacturing. The failure to adopt new technologies has already decreased the international cost competitiveness and production efficiency of U.S. industry, thus causing unemployment.

Gold, Bela, Gerhard Rosegger and Myles G. Boylan, Jr. *Evaluating Technological Innovations.* D.C. Heath and Company, Toronto, Canada, 1980.

The authors attempt to establish an improved analytical foundation for firms to evaluate new technologies. The empirical data is taken from the iron and steel industry. One of the authors' many conclusions is that successful technological innovation by one firm in an industry is not evidence that other firms in the same industry should or will promptly adopt the same innovation.

Goldstein, Harvey. *Occupational Employment Projections for Labor Market Areas: An Analysis of Alternative Approaches.* U.S. Department of Labor, R&D Monograph 80, U.S. Government Printing Office, Washington, DC, 1981.
This monograph describes the general features of the OES approach to occupational employment projections and examines the potential of econometric models and input-output models for improving projections. Local occupational forecasting requires improved local labor market data and analysis.

Gordus, Jeanne P., Paul Jarley and Louis A. Ferman. *Plant Closings and Economic Dislocation.* The W.E. Upjohn Institute for Employment Research, Kalamazoo, MI, 1981.
Presents an overview of 20 plant closing studies published in the last two decades. Emphasizes what we know about plant closings and what research remains to be done.

Grabbe, Eugene M. and Donald L. Pyke. "An Evaluation of the Forecasting of Information Processing Technology and Applications," *Technological Forecasting and Social Change,* Vol. 4, 1972, pp. 143-150.
The Delphi method of forecasting is described and evaluated by comparing events forecasted with actual dates of occurrence. Although the data provided is inconclusive, the indication is that information processing technology and application are advancing more rapidly than expected.

"Growth Industries of the Future," *Newsweek,* October 12, 1982, p. 82.
Discusses the different forecasts of the future job markets. Robotics will be one of the major growth areas.

Haber, William, Louis A. Ferman and James R. Hudson. *The Impact of Technological Change.* The W.E. Upjohn Institute for Employment Research, Kalamazoo, MI, 1963.
One of the early reviews of the empirical research on the impact of technological change. It specifically assesses the evidence of job displacement by reviewing 17 studies conducted between 1929 and 1961.

Hekman, J. S. "The Future of High Technology Industry in New England: A Case Study of Computers," *New England Economic Review,* January/February 1980, pp. 5-17.

Discusses the computer industry, including main manufacturing centers and branches of these centers. Suggests that the smaller branches are more mobile and will be indicators of growth and mobility in terms of geographical location. The South Atlantic, Southwest and Pacific regions tend to be most attractive for computer branch locations, while New England is a strong location for new firms with new products.

———. "Can New England Hold onto Its High Technology Industry?" *New England Economic Review,* March/April 1980, pp. 35-44.

The author examines the medical instrument industry and concludes that New England will retain its attraction for these firms.

Hollomon, Herbert J. *Technical Change and American Enterprise.* Report No. 9, National Planning Association, Washington, DC, 1974.

Discusses the process of technological change and makes recommendations for private and public policies. Factors considered include utilization of existing knowledge toward new technology, increased support of applied sciences and engineering, the need for collective R&D, business and the government's role in protecting the consumer, and policy changes necessary to insure that the negative impacts of technological change are not absorbed only by the individual workers and firms involved.

Hunt, H. Allan and Timothy L. Hunt. *Robotics: Human Resource Implications for Michigan.* The W. E. Upjohn Institute for Employment Research, Kalamazoo, MI, 1982a.

Final report to the Michigan Occupational Information Coordinating Committee (MOICC) regarding the impacts of robotics on the State of Michigan.

———. *Robotics: Human Resource Implications for Michigan, A Summary.* Michigan Occupation Information Coordinating Committee, 1982b.

Summarizes the findings of the full state report.

Industrial Technology Institute. *The Program for the Industrial Technology Institute.* Preliminary draft proposal, Industrial Technology Institute, Ann Arbor, MI, August 1982. Mimeographed.

Outlines the goals and objectives of the Industrial Technology Institute as well as the role it will play in research and development.

Institute of Science and Technology, University of Michigan. *Automatic Factory Opportunities in Michigan: I. Robotics.* January 1982.
Analyzes the robotics industry and its potential in the State of Michigan from an economic development perspective. Topics include robotics research and training programs in Michigan, short descriptions of robotics manufacturers in Michigan, the Michigan business environment, and business services provided by the Office of Economic Development (OED), Michigan Department of Commerce.

Jablonowski, Joseph. "Robots: Looking Over the Specifications," *American Machinist,* Special Report 745, May 1982, pp. 163-178.
Discusses the specifications, capabilities, and costs of industrial robots displayed at Robotics VI Conference (March 1982) by firm and model name.

Kendrick, John W. "The Coming Rebound in Productivity," *Fortune,* June 28, 1982, pp. 25-28.
Explains that productivity is partially based on the level of business activity. Productivity will improve over the next several years due to several forces including the Economic Recovery Act of 1981, increased awareness and confrontation of productivity problems, expanded investment in plants and equipment and foreign competition.

Krause, Jeffrey M. "Robotics Impact on Human Resources." Unpublished thesis presented to General Motors Institute. Detroit, MI, April 1982.
Given the goal at GM for increased use of advanced technology systems, especially robots, the author assesses the implications of robotics on human resources emphasizing that successful integration of robots at GM can lead to efficiency and productivity that will benefit GM, the workers and the community.

Leon, Carol Boyd. "Occupational Winners and Losers: Who They Were During 1972-80," *Monthly Labor Review,* June 1982, pp. 18-28.
Reports changes in employment between 1972 and 1980 based on the Current Population Survey. Includes the occupations that experienced significant growth or decline.

Leontief, Wassily W. "The Distribution of Work and Income," *Scientific American,* September 1982, pp. 188-204.
Discusses the likely impact of new technologies on employment and income in the American economy by the year 2000. Suggests work sharing and other measures to insure an equitable distribution of income.

Levitan, Sar A. and Clifford M. Johnson. *Second Thoughts on Work.* The W. E. Upjohn Institute for Employment Research, Kalamazoo, MI, 1982.

Among other things, this monograph includes a section on the likely effects of robots on the workforce and workplace. Concludes that there are limits to the rate of technological change and also questions several popular perceptions about economic growth and technological change.

"Little Corporate Zest for Leading a Recovery," *Business Week,* December 13, 1982, p. 14.

Based on a survey conducted by Louis Harris and Associates, the majority of executives report that tactics to improve business activity in a flat economy include, among other things, increased output without significantly adding to the workforce, increased output without additional capital financing, and expanding inventories based only on demand and output.

Lund, Robert T., Christopher J. Barnett and Richard M. Kutta. *Numerically Controlled Machine Tools and Group Technology: A Study of U.S. Experience.* Center for Policy Alternatives, MIT, Cambridge, MA, January 1978.

This study examines the experience of N/C machine tools as the forerunner of new computer-based manufacturing and the effects of this technology on the discrete product manufacturing industry. The question this study seeks to answer is "whether new manufacturing technologies could bring the economics of batch manufacturing sufficiently close to those of large-scale production to make smaller-scale manufacturing more attractive."

Luria, Daniel. "Technology, Employment and the Factory of the Future." Presented to the SME Autofact III Conference, Detroit, November 9, 1981.

This study suggests that the U.S. auto industry must automate, rebuild and retool or lose even more markets to the Japanese. However, the U.S. must plan for these changes. Luria suggests reducing the length of the workweek, increasing international communication and support among unions, requiring advance notification of plant closures, reevaluating traditional management-union relationships, retraining of displaced workers and increasing job security for workers.

Lustgarten, Eli S. "Robotics and Its Relationship to the Automated Factory," in *Exploratory Workshop on the Social Impacts of Robotics: Summary and Issues.* Office of Technology Assessment, U.S. Government Printing Office, Washington, DC, February 1982, pp. 119-36.
Lustgarten, an investment analyst for Paine Webber Mitchell Hutchins, Inc., projects a $2.0 billion U.S. market for the robotics industry by 1990. The adoption of robots will be spurred in part by the aging of U.S. plant and equipment and the expected mid-1980s drop in the entry level workforce.

Macut, John J. "New Technology in Metalworking," *Occupational Outlook Quarterly,* February 1965, pp. 1-6.
Discusses the use of N/C machine tools compared to the conventional machine tools in terms of productivity, efficiency and quality control and the effects of this technology on the workforce.

Malecki, Edward. "Product Cycles, Innovation Cycles, and Regional Economic Change," *Technological Forecasting and Social Change,* May 1981, pp. 309-323.
This review paper examines the implications of technological change for regional economic development and policy from the perspective of product cycles and innovation cycles. The review is fairly comprehensive and includes a lengthy listing of references.

————. "Public and Private Sector Interrelationships, Technological Change, and Regional Development," *Papers of The Regional Science Association,* Volume 47, 1981, pp. 121-137.
Reviews the available evidence of the influence of public and private sector research and development spending on regional economic development. The study concludes that private research and development tends to lead rather than follow government and research activity.

Mansfield, Edwin. *Industrial Research and Technological Innovation: An Econometric Analysis.* W.W. Norton and Company, Inc., New York, 1968.
This book brings together some of the results of Mansfield's quantitative studies of technological change and innovation.

Mansfield, Edwin, et al. *The Production and Application of New Industrial Technology.* W.W. Norton and Company, Inc., New York, 1977.

Mansfield compares private and social rates of return for new technological innovations and concludes that in many cases the social returns are much higher than the private returns.

Mansfield, Edwin, et al. *Research and Innovation in the Modern Corporation.* W.W. Norton and Company, Inc., New York, 1971a.
One of the pioneering efforts by Mansfield and his students. It is impossible to adequately summarize this work, but one of the major insights is that it took a decade or more for a majority of the firms to adopt a specific new process technology.

Mansfield, Edwin. *Technological Change.* W.W. Norton and Company, Inc., New York, 1971b.
Surveys the literature of technological change. This book provides a concise introduction to Mansfield's voluminous work as well as others. The review is relatively nontechnical and written specifically for a general audience. There is also a balanced presentation of the relevant policy issues.

Martin, Gail M. "Industrial Robots Join the Workforce," *Occupational Outlook Quarterly,* Fall 1982a, pp. 2-11.
Describes the types and uses of robots currently and addresses concerns over present and future trends in the robotics industry, particularly those related to the effects this industry will have on the workforce and society.

————. "Manufacturing Engineering," *Occupational Outlook Quarterly,* Fall 1982b, pp. 22-26.
Describes the changing role of the manufacturing engineer due to the introduction of robotics and CAD/CAM and new concerns such as energy costs, competition of foreign manufacturers and lagging productivity.

Meisner, Charlotte. *High Technology Employment: Massachusetts and Selected States 1975-1981.* Massachusetts Division of Employment Security, Job Market Research, 1982.
High technology industries are characterized by a high ratio of research and development to sales, high value-added products, high ratio of scientists and engineers, and high growth rates. This somewhat broad definition is operationalized by selecting 20 industries at the 3-digit level from the SIC system. The study then presents comparative data for the states and industries selected.

Michigan Employment Security Commission, Occupational Employment Statistics Unit. *Michigan Occupational Employment Statistics*

for Manufacturing Industries. Michigan Employment Security Commission, Detroit, April 1981a.
Reports results of the Occupational Employment Survey (OES) for Michigan manufacturing industries in 1977 at the 2-digit SIC code level of industrial detail.

Michigan Employment Security Commission, Bureau of Research and Statistics. *Motor Vehicle and Related Industries in Michigan.* Michigan Employment Security Commission, Detroit, Summer 1981b.
The results of a survey of all manufacturing industries directly involved in supplying parts, materials and special tool and dies for the automobile industry are discussed revealing that 55 percent of Michigan's manufacturing sector is employed in the automobile or automobile-related industries.

———. *Occupational Supply and Demand in Michigan.* Michigan Employment Security Commission, Detroit, Winter 1982.
This pioneering effort funded by the Michigan Occupational Information Coordinating Committee is a first attempt to bring together the information available about the demand and supply for various occupations in the State of Michigan. It includes analyses of 60 occupational clusters primarily accessed through the vocational education system.

Milliken, William G. *A Plan to Increase the High Technology Component of Michigan's Economy.* September 1981a.
Outlines goals and objectives to increase the high technology component of Michigan's economy describing the accomplishments of the High Technology Task Force and recommendations for future activities.

———. *Special Message to the Michigan Legislature on Economic Development.* September 17, 1981b.
Presents Governor Milliken's plan for the future economic development of the State of Michigan.

"Motor Vehicles, Model Year 1982," *Survey of Current Business,* October 1982, pp. 20-23.
Describes 1982 as the worst year since 1961 for the motor vehicle industry. The economic and financial conditions of the major auto manufacturers continued to deteriorate.

Nabseth, Lars and George F. Ray, eds. *The Diffusion of New Industrial Processes.* Cambridge University Press, London, 1974.

An international study of technological diffusion that spans six years and six research institutes. The processes analyzed included numerically controlled machine tools. Even though numerical control appeared appropriate for smaller firms which produce small batch jobs, it was not adopted to any significant degree because of the large initial financial cost.

National Center for Productivity and Quality of Working Life. *New Technologies and Training in Metalworking.* Washington, DC, 1978.
Analyzes the development of numerical control and other technologies in metalworking.

National Center for Productivity and Quality of Working Life. *Productivity and Job Security: Retraining to Adapt to Technological Change.* Washington, DC, 1977.
Presents five case studies on worker retraining to determine the advantages and disadvantages of each and their effectiveness in protecting job security. Policy implications are discussed regarding management's role in incorporating new technology to achieve higher productivity and maintain competitiveness while insuring the job security of its workers through retraining.

National Commission on Technology, Automation, and Economic Progress. *Technology and the American Economy.* Volume 1. Prepared for the U.S. Congress, 1966.
This national commission assesses the impacts of technological change and policy challenges resulting from it. While the Commission found that fears of widespread unemployment due to automation were not well founded, they did find reason to be concerned about adjustments to technological change and its impact on particular groups and sectors of the economy. Contains a set of recommendations for a comprehensive program addressing the needs of those workers displaced by technological change.

National Science Foundation. *Science and Engineering Degrees: 1950-80.* National Science Foundation, Special Report NSF-32-307, Washington, DC, 1982a.
A statistical data source that provides detailed estimates of science and engineering degree production.

————. "Labor Markets for New Science and Engineering graduates in Private Industry," *Science Resource Studies Highlights,* NSF 82-310, Washington, DC, June 9, 1982b, pp. 1-5.

Presents data on labor market conditions for science and engineering graduates based on a survey of 255 firms in 1981.

———. "Engineering Colleges Report 10% of Faculty Positions Vacant in Fall of 1980," *Science Resource Studies Highlights,* National Science Foundation, NSF 81-322, Washington, DC, November 1981a, pp. 1-4.
Reports on the results of a survey of 181 engineering colleges.

———. "Science and Engineering Faculty with Recent Doctorates Fell to One-fifth of Total in 1980," *Science Resource Studies Highlights,* National Science Foundation, NSF 81-318, Washington DC, October 1981b, pp. 1-4.
Reports the findings of a 1980 survey supported by the National Science Foundation.

———. *National Patterns of Science and Technology Resources 1981.* U.S. Government Printing Office, NSF 81-311, Washington, DC, April 1981c.
A statistical compilation of U.S. research and development resources. Includes an evaluation of labor market conditions for science and engineering personnel.

———. *Science and Engineering Employment: 1970-80.* National Science Foundation, Special Report NSF 81-310, Washington, DC, March 1981.
Charts growth in science and engineering employment levels over the decade of the 70s. Includes data on distribution of scientists and engineers by sector and specialty. Also reports indices for research and development employment and total funding 1970 to 1980.

———. *Problems of Small High-Technology Firms.* National Science Foundation, Special Report NSF 81-305, Washington, DC, December 1981.
Results of a survey of 1232 high-tech firms with less than 500 employees. Major problems are identified in the financial, personnel and government relations areas. No policy implications are offered.

Nelson, Richard R., Merton J. Peck and Edward D. Kalacheck. *Technology, Economic Growth and Public Policy.* The Brookings Institution, Washington, DC, 1967.
This work draws together most of the previous research on technological progress. Topics include the way in which the economy adjusts to technological change and possible public policy concerns.

The authors propose a framework for delineating the roles of private, public and governmental financing of R&D.

Obrzut, John A. "Robotics Extends A Helping Hand," *Iron Age,* March 19, 1982, pp. 59-83.
Suggests that the robot industry finds itself popularized in the media but relatively short on orders in 1982.

Porter & Novelli Associates. *Targeted Industry Marketing Program: Advanced Manufacturing Systems Report.* Mimeographed. Michigan State Department of Commerce, 1981.
Assesses the level of interest as well as the locational factors that attract business to Michigan by interviewing firms in and out of state.

Prab Robots, Inc. Annual Report, 1981. Kalamazoo, Michigan.
Presents complete operating results for this firm for the fiscal year ending October 31, 1981.

Putnam, George P. "Why More NC Isn't Being Used," *Machine and Tool Blue Book,* September, 1978, pp. 98-107.
Reports results of a survey of small machine tool firms who were considered candidates for use of numerical control. Found that 72 percent of the firms had not formally evaluated the applicability of numerical control for their firms.

Rees, John. "Technological Change and Regional Shifts in American Manufacturing," *Professional Geographer,* Vol. 31, No. 1, 1979, pp. 45-54.
Discusses the changes taking place in the American economic system and the effects that these changes are having on the industrial geography of the United States.

Riche, W. Richard. "Impact of Technological Change." Mimeographed. Prepared for the U.S. Department of Labor for presentation at the *Organization for Economic Cooperation and Development's Second Special Session on Information Technologies, Productivity, and Employment* in Paris, France, October 1981.
Suggests that American business, workers, and consumers have shared the benefits of technological change and will likely continue to do so in the future. Adoption of new technologies has not resulted in layoffs of workers because the private sector has retrained and reassigned displaced workers or accomplished employment reductions through normal attrition. A short section about robotics notes that widespread adoption likely awaits the development of satisfactory sensing devices.

"Robotics Class Looks Ahead," *The Detroit News,* March 10, 1982.
Reports on the Warren, Michigan robotics technician training program conducted under the Comprehensive Employment and Training Act (CETA).

Robot Institute of America. *RIA Worldwide Survey and Directory on Industrial Robots.* Robot Institute of America, Dearborn, MI, 1981.
Reports the results of a survey on the use of robots in 18 countries. The survey answers questions regarding the various applications of robots, current population, financial information and future trends of robotics. Includes a directory of the leading manufacturers, distributors, component suppliers and research and government organizations involved in the field.

"A Robotics Mecca in Michigan? Car Sales Must Rebound First," *Detroit Free Press,* October 11, 1982.
Reports that lagging auto sales have delayed some robot purchases by the auto firms and suggests that the future of the robot industry in Michigan depends in part on a recovery in auto sales. Reports also that many of the new entrants in the robot market have had little or no sales in 1982.

Rosenthal, Neal H. "Shortages of Machinists: An Evaluation of the Information," *Monthly Labor Review,* July 1982, pp. 31-36.
Presents an interesting discussion of what can be gleaned from various data sources about the current situation and future outlook for machinists. The data are consistent with a shortage, but sufficient information is not available to quantify that shortage.

Ruben, George. "Developments in Industrial Relations," *Monthly Labor Review,* September 1982a, pp. 44-45.
Discusses the job security focus of the new collective bargaining agreement at General Electric Company.

———. "Developments in Industrial Relations," *Monthly Labor Review,* October 1982b, p. 44.
Discusses the new collective bargaining agreement for Westinghouse Electric Corporation.

Rumberger, Russell W. "The Changing Skill Requirements of Jobs in the U.S. Economy," *Industrial and Labor Relations Review,* Vol. 34, No. 4, July 1981, pp. 578-590.
This study measures the shifts in job skill requirements in the U.S. economy from 1960 to 1976. Both changes in the distribution of

employment among occupations and changes in the skill requirements of individual occupations are addressed.

Russo, G. Paul. "Robotics at Chrysler," in *Robotics and the Factory of the Future,* presented at the University of Michigan Management Briefing Seminars, Traverse City, MI, August 2, 1982.
Chrysler is using 240 robots today, most in welding applications. By the end of 1988 Chrysler expects to have 987 robots installed in their plants, a growth rate of approximately 30 percent.

Russell, Jack. "Michigan's Ailing Economy: Is Robotics the Cure?" Mimeographed.
The potential for the emerging robotics industry in the State of Michigan is discussed. He concludes that robotics has little employment potential for the state. Russell goes on to provide an alternative possibility for economic growth—that of energy hardware—which he feels would better utilize the already existing metalworking industry and provide 100,000 industrial jobs.

Sahal, Devendra. *Patterns of Technological Innovation.* Addison-Wesley Publishing Company, Inc., Reading, MA, 1981.
Sahal proposes (and supports with numerous case studies) an evolutionary concept of technological innovation which is more eclectic than traditional economic approaches. In his own words, ". . .technological innovation is too significant a process to be left to economists and engineers. What is needed is an independent science of technology. My attempt in this book has been to provide the essentials of this emerging science." Needless to say, the book is novel, complex and rather comprehensive.

Schreiber, Rita R. "Meeting the Demand for Robotics Technicians," in *Robotics Today,* Summer 1981, reprinted in *Robotics Today '82 Annual Edition,* Society of Manufacturing Engineers, Dearborn, MI, 1982, pp. 78-79.
Describes the development of the first robotics technician curricula in the U.S.: Macomb Community College, Warren, MI in 1978.

Smith, Donald N. and Richard C. Wilson. *Industrial Robots: A Delphi Forecast of Markets and Technology.* Society of Manufacturing Engineers, Dearborn, MI, 1982.
Reports results of a Delphi survey on many technical, marketing and sociological aspects of the development of industrial robots. Over 200 questions were asked in round one, while rounds two and three

repeated some questions of round one as well as adding supplemental questions suggested by the panel of experts.

Smith, Donald N., Peter G. Heytler and Murray D. Wikol. "Sociological Effects of the Introduction of Robots in U.S. Manufacturing Industry." Industrial Development Division, Institute of Science and Technology, University of Michigan, Ann Arbor, MI. Paper presented at the *CAMPRO '82 Conference on Computer Aided Manufacturing and Productivity,* October 1982.
Discusses the implications of robotics in terms of current and future developments. According to the authors, the market capacity for the industry is expected to "burgeon" into a multi-billion dollar industry by 1990. The potential displacement resulting from the growth of the industry is expected to be offset by retraining.

Sternlieb, George and James W. Hughes, eds. *Post-Industrial America: Metropolitan Decline and Inter-Regional Job Shifts.* Center for Urban Policy Research, Rutgers—The State University of New Jersey, New Brunswick, NJ, 1975.
A collection of papers on the plight and future of America's cities.

Tanner, William R., ed. *Industrial Robots Volume 2/Applications.* Robotics International of SME, Society of Manufacturing Engineers, Dearborn, MI, 1981.
A collection of journal articles, technical papers and proceedings outlining the most recent technological advancements in robotics. The materials in this volume cover the use of robotics in the areas of Material Handling, Machine Loading, Die Casting, Press Loading, Forging and Heat Treating, Foundries, Plastics Molding, Welding, Assembly and other areas as well. The cost-effectiveness and productivity implications are discussed in each chapter.

Tanner, William R. and William F. Adolfson. *Robotics Use in Motor Vehicle Manufacture.* Report to the U.S. Department of Transportation, February 1982.
Discusses the expanding utilization of robots in the automobile industry predicting that by 1990 the robot population could be 35,000 or more. Robot use will expand because of their effectiveness in improving productivity and product quality.

Terleckyj, Nestor E. and Martin K. Holdrich. *Sectoral Growth in Output, Productivity and Employment, 1981-2000.* National Planning Association, Report No. 81-N-2, Washington, DC, March 1982.
Projects growth in output and employment for broad sectors of the

U.S. economy. Forecasts that employment in manufacturing is expected to decline by 3.5 million in the 1980s and 1990s and continued growth is expected in employment in trades, finance and service sectors.

United Auto Workers. *Technology: Promises and Problems.* A Policy Statement, October 1981.
Discusses new technology and the UAW's policies regarding worker protection and job security. It also assesses the role of public policy in the same context.

U.S. Congress, Congressional Budget Office. *Dislocated Workers: Issues and Federal Options.* U.S. Government Printing Office, Washington, DC, July 1982.
Discusses the underlying causes and impacts of worker dislocation and analyzes the possibilities for federal aid to these workers.

———. *Location of High Technology Firms and Regional Economic Development.* U.S. Government Printing Office, Washington, DC, 1982.
A staff study prepared for the Subcommittee on Monetary and Fiscal Policy. Emphasizes the importance of high technology industry for economic growth in the U.S. The results are based on a survey of 691 high technology firms in an attempt to gain more knowledge regarding the locational decision making factors of high technology firms. Factors identified in the study include availability of skilled labor, labor costs, state and local taxes and proximity of educational institutions. Relative growth of high technology industries will be fastest in the Midwest.

———. *U.S. Economic Growth from 1976 to 1986: Prospects, Problems and Patterns,* Vol. 9, *Technological Change.* U.S. Government Printing Office, Washington, DC, 1977.
Indicates that a stagnant or slow growing economy with high levels of unemployment is not the appropriate environment in which innovative activity will flourish. The study cautions policymakers not to place excessive expectations on technological change as a solution for economic woes.

U.S. Congress. "Robotics: Economic and Social Implications." Congressional Clearinghouse on the Future, 1981.
Key individuals such as Mr. Stanley Polcyn, V.P., Unimation, Dr. Angel Jordan, Dean, Carnegie Institute of Technology, Richard Beecher, General Motors Corporation, William Spurgeon, National

Science Foundation and Thomas Weekley from UAW, discuss robotics including its present and future implications for society and the economy.

U.S. Congress, Office of Technology Assessment. *Exploratory Workshop on the Social Impacts of Robotics: Summary and Issues.* U.S. Government Printing Office, Washington, DC, July 1981.
This background paper summarizes the results of an exploratory workshop designed to examine the state of robotic's technology and possible public policy issues. It also includes four separate papers that were used as starting points for the workshop participants. The papers by Albus, Aron, Gold, and Lustgarten are entered separately in this bibliography. The participants at the workshop agreed that robotics was only one part of the technological base designed to increase industrial automation; the robot market is in its infancy and there is a shortage of trained technical experts in robotics. Not surprisingly, there was less agreement on the social and economic impacts, although most experts argue that new technology creates more jobs than are eliminated, and historically that certainly has been the case.

U.S. Department of Commerce, Bureau of the Census. *1977 Census of Manufactures.* U.S. Government Printing Office, Washington, DC, 1981a.
This document is the most comprehensive source of data for U.S. manufacturing industries.

U.S. Department of Commerce. *United States Automobile Industry Status Report.* Submitted to the U.S. Senate Committee on Finance, Subcommittee on International Trade, December 1, 1981b.
The current slump in the automobile industry is more than cyclical. Poor sales also reflect a downward trend in demand for automobiles due to international competition and demographic factors. The result has been the erosion of financial strength for the U.S. auto industry.

U.S. Department of Labor, Bureau of Labor Statistics. *Technology and Labor in Four Industries.* Bulletin 2104, January 1982a.
Reviews major technological changes among the following selected industries: meat products, foundries, metalworking machinery and electrical and electronic equipment. Discusses the effects these changes will have on productivity and occupations during the next 5-10 years.

———. *Employment and Earnings,* Vol. 29, No. 3, March 1982b. U.S. Government Printing Office, Washington, DC.

A comprehensive source of earnings and employment data for the U.S. and the individual states.

———. *Occupational Outlook Handbook, 1982-83 Edition.* Bulletin 2200, U.S. Government Printing Office, Washington, DC, April 1982c.
The primary source of information about specific occupations and the outlook for employment in those occupations. Very useful for vocational guidance. The biennial volume is the major focus of BLS occupational forecasting efforts.

———. *Projected Occupational Staffing Patterns of Industries.* OES Technical Paper No. 2, March 1981a.
This paper describes in considerable technical detail the methods developed by BLS to project future occupational employment by industry. Also provides a comparative analysis of different techniques for projecting future staffing patterns based on 1973, 1975 and 1978 OES nonmanufacturing surveys.

———. *Comparison of Occupational Employment in the 1978 Census-Based and OES Survey-Based Matrices.* Technical Paper No. 1, April 1981b. Mimeographed.
This technical paper compares the levels of employment for 1978 from the census-based household survey and the OES employer survey. While the results are similar in the two surveys, there proved to be large variation in some specific occupations.

———. *Productivity and the Economy: A Chartbook.* Bulletin 2084, U.S. Government Printing Office, Washington, DC, October 1981c.
Presents data on various measures of U.S. productivity.

———. *Occupational Employment in Manufacturing Industries, 1977.* Bulletin 2057, U.S. Government Printing Office, Washington, DC, March 1980.
Reports the results of the 1977 OES survey for manufacturing industries. A detailed occupational profile is provided for industries at the 2-digit level. Also shows changes in proportions of broad occupational groups since 1971.

———. *Tomorrow's Manpower Needs.* Vol. 1-4, Bulletin 1606, U.S. Government Printing Office, Washington, DC, 1969.
This four volume set reports on the initial BLS effort to develop consistent and detailed projections of employment by occupation for the nation and local areas. It is the forerunner of the current OES projection system.

———. *Occupational Employment Patterns for 1960 and 1975.* Bulletin 1599, U.S. Government Printing Office, Washington, DC, 1968.
This report projected the occupational employment patterns by industry to 1975. It was essentially based on the BLS Industry-Occupational Employment Matrix for 1960 and employment trends revealed from 1950 to 1960 in the Current Population Survey and the BLS Industry Employment Statistics series.

U.S. General Accounting Office. *Advances in Automation Prompt Concerns Over Increased U.S. Unemployment.* U.S. General Accounting Office, Report No. AFMD-82-44, May 1982.
Survey of attitudes on the employment impacts of automation. Explains the reasons for differing opinions on the long-run employment impact of automation. Points out that we know little about the short-term consequences of displacement.

Vedder, Richard. *Robotics and the Economy.* Prepared for the Subcommittee on Monetary and Fiscal Policy of the Joint Economic Committee, Congress of the United States, U.S. Government Printing Office, Washington, DC, 1982.
This study examines the growth of the robotics industry and its impacts on employment. Robots will be introduced gradually, so the majority of displaced workers will be spared unemployment through retirement of some workers or retraining. In fact robots may have a positive net effect on employment because it will spur economic growth. Retraining issues will be addressed through collective bargaining and modification and expansion of vocational education.

Verway, David I. "Michigan Outlook," in *The Michigan Economy.* Vol. 1, No. 1, June 1982.
Presents an analysis of the current status of the Michigan economy and likely prospects for the future of the state.

Wisnosky, Dennis E. "On the Importance of Engineered Solutions," *Robotics Today,* June 1982, p. 22.
Assesses the future for robotics technology. Wisnosky concludes that the automatic factory is some years in the future because of a lack of data-based technology, among other factors.